Lecture Notes in Bioinformatics 6254

Edited by S. Istrail, P. Pevzner, and M. Waterman

Subseries of Lecture Notes in Computer Science

Patrick Lambrix   Graham Kemp (Eds.)

Data Integration in the Life Sciences

7th International Conference, DILS 2010
Gothenburg, Sweden, August 25-27, 2010
Proceedings

 Springer

Series Editors

Sorin Istrail, Brown University, Providence, RI, USA
Pavel Pevzner, University of California, San Diego, CA, USA
Michael Waterman, University of Southern California, Los Angeles, CA, USA

Volume Editors

Patrick Lambrix
Linköpings universitet
Department of Computer and Information Science, 581 83 Linköping, Sweden
E-mail: patla@ida.liu.se

Graham Kemp
Chalmers University of Technology and University of Gothenburg
Computer Science and Engineering, 412 96, Gothenburg, Sweden
E-mail: kemp@chalmers.se

Library of Congress Control Number: 2010932135

CR Subject Classification (1998): H.3, J.3, I.2, H.4, C.2, H.5

LNCS Sublibrary: SL 8 – Bioinformatics

ISSN   0302-9743
ISBN-10   3-642-15119-1 Springer Berlin Heidelberg New York
ISBN-13   978-3-642-15119-4 Springer Berlin Heidelberg New York

springer.com

© Springer-Verlag Berlin Heidelberg 2010
Printed in Germany

Typesetting: Camera-ready by author, data conversion by Scientific Publishing Services, Chennai, India
Printed on acid-free paper   06/3180

Preface

The development and increasingly widespread deployment of high-throughput experimental methods in the life sciences is giving rise to numerous large, complex and valuable data resources. This foundation of experimental data underpins the systematic study of organisms and diseases, which increasingly depends on the development of models of biological systems. The development of these models often requires integration of diverse experimental data resources; once constructed, the models themselves become data and present new integration challenges for tasks such as interpretation, validation and comparison.

The Data Integration in the Life Sciences (DILS) Conference series brings together data and knowledge management researchers from the computer science research community with bioinformaticians and computational biologists, to improve the understanding of how emerging data integration techniques can address requirements identified in the life sciences.

DILS 2010 was the seventh event in the series and was held in Gothenburg, Sweden during August 25–27, 2010. The associated proceedings contain 14 peer-reviewed papers and 2 invited papers. The sessions addressed *ontology engineering*, and in particular, evolution, matching and debugging of ontologies, a key component for semantic integration; *Web services* as an important technology for data integration in the life sciences; *data and text mining* techniques for discovering and recognizing biomedical entities and relationships between these entities; and *information management*, introducing data integration solutions for different types of applications related to cancer, systems biology and microarray experimental data, and an approach for integrating ranked data in the life sciences. The invited talk by Juliana Freire reviewed state-of-the-art techniques, research challenges and open problems involved in managing provenance throughout the data life cycle. Joel Saltz discussed the requirements for and design of a system for composing, executing and exploring in silico experiments involving microscopy images.

The editors would like to thank the Program Committee and the external reviewers for their work in enabling the timely selection of papers for inclusion in the proceedings. We acknowledge the support of AstraZeneca as well as of Chalmers University of Technology and Linköping University. We also appreciate our cooperation with EasyChair as well as our publisher Springer.

June 2010

Patrick Lambrix
Graham Kemp

Organization

General Chairs

Graham Kemp   Chalmers University of Technology, Sweden
Patrick Lambrix   Linköping University, Sweden

Local Organization

Rebecca Cyrén   Chalmers University of Technology, Sweden
Devdatt Dubhashi   Chalmers University of Technology, Sweden
Merja Karjalainen   Chalmers University of Technology, Sweden
Anna-Lena Karlsson   Chalmers University of Technology, Sweden
Graham Kemp   Chalmers University of Technology, Sweden
Caroline Olsson   Chalmers University of Technology, Sweden

Program Chairs

Patrick Lambrix   Linköping University, Sweden
Graham Kemp   Chalmers University of Technology, Sweden

Program Committee

Hans Åhlfeldt   Linköping University, Sweden
Chris Baker   University of New Brunswick, Canada
Albert Burger   Heriot-Watt University and MRC Human
     Genetics Unit, UK
Sarah Cohen-Boulakia   Université Paris-Sud, France
Terence Critchlow   Pacific Northwest National Laboratory, USA
Christine Froidevaux   Université Paris-Sud, France
Carole Goble   University of Manchester, UK
Peter Karp   SRI International, USA
Graham Kemp   Chalmers University of Technology, Sweden
Jessie Kennedy   Edinburgh Napier University, UK
Patrick Lambrix   Linköping University, Sweden
Mong Li Lee   National University of Singapore, Singapore
Ulf Leser   Humboldt-Universität zu Berlin, Germany
Frédérique Lisacek   Swiss Institute of Bioinformatics, Switzerland
Bertram Ludäscher   University of California, Davis, USA
Marco Masseroli   Politecnico di Milano, Italy
Paolo Missier   University of Manchester, UK
See-Kiong Ng   Institute for Infocomm Research, Singapore

José Luis Oliveira	Universidade de Aveiro, Portugal
Alexandra Poulovassilis	Birkbeck College, UK
Erhard Rahm	Universität Leipzig, Germany
Louiqa Raschid	University of Maryland, USA
Christopher J. Rawlings	Rothamsted Research, UK
Falk Schreiber	Martin Luther University Halle-Wittenberg, Germany
Michael Schroeder	TU Dresden, Germany
Christopher Southan	AstraZeneca, Sweden
Chris Stoeckert	University of Pennsylvania, USA
Lena Strömbäck	Linköping University, Sweden
Mark Wilkinson	University of British Columbia, Canada
Anil Wipat	University of Newcastle, UK

Additional Reviewers

Dimitra Alexopoulou
Jonas Bergman Laurila
Julie Bernauer
Heiko Dietze
Anika Groß
Toralf Kirsten
Arash Shaban-Nejad
Karen Sutherland
Thomas Wächter

Table of Contents

Information Management

Provenance Management for Data Exploration

Juliana Freire

SCI Institute, University of Utah, USA
Linköping University, Sweden

Computing has been an enormous accelerator to science and industry alike and it has led to an information explosion in many different fields. The unprecedented volume of data acquired by sensors, derived by simulations and analysis processes, and shared on the Web opens up new opportunities, but it also creates many challenges when it comes to managing and analyzing these data. In this talk, I discuss the importance of maintaining detailed provenance (also referred to as lineage and pedigree) for digital data. Provenance provides important documentation that is key to preserve data, to determine the data's quality and authorship, to understand, reproduce, as well as validate results [9,4,3]. I will review some of the state-of-the-art techniques, as well as research challenges and open problems involved in managing provenance throughout the data life cycle [28,11,17,27,24,6,7,13,12]. I will also discuss benefits of provenance that go beyond reproducibility, including techniques and tools we have developed that leverage provenance information to support reflective reasoning and collaborative data exploration and visualization [11,1,25,23,15,22,5,21,14,19,20]. I conclude with a discussion on new applications that are enabled by provenance. In particular, I present how provenance can be used to aid in teaching [26], to create reproducible papers [16,8], and as the basis for social data analysis [10,18,2].

References

1. Callahan, S., Freire, J., Santos, E., Scheidegger, C., Silva, C., Vo, H.: Managing the evolution of dataflows with vistrails (Extended Abstract). In: IEEE Workshop on Workflow and Data Flow for Scientific Applications, SciFlow (2006)
2. CrowdLabs, http://www.crowdlabs.org
3. Davidson, S.B., Boulakia, S.C., Eyal, A., Ludäscher, B., McPhillips, T.M., Bowers, S., Anand, M.K., Freire, J.: Provenance in scientific workflow systems. IEEE Data Eng. Bull. 30(4), 44–50 (2007)
4. Davidson, S.B., Freire, J.: Provenance and scientific workflows: challenges and opportunities. In: SIGMOD, pp. 1345–1350 (2008)
5. Ellkvist, T., Koop, D., Anderson, E.W., Freire, J., Silva, C.T.: Using provenance to support real-time collaborative design of workflows. In: Freire, J., Koop, D., Moreau, L. (eds.) IPAW 2008. LNCS, vol. 5272, pp. 266–279. Springer, Heidelberg (2008)
6. Ellkvist, T., Strömbäck, L., Lins, L.D., Freire, J.: A first study on strategies for generating workflow snippets. In: Proceedings of the ACM SIGMOD International Workshop on Keyword Search on Structured Data (KEYS), pp. 15–20 (2009)
7. Ellqvist, T., Koop, D., Freire, J., Silva, C., Stromback, L.: Using mediation to achieve provenance interoperability. In: IEEE Congress on Services, pp. 291–298 (2009)
8. Fomel, S., Claerbout, J.F.: Guest editors' introduction: Reproducible research. Computing in Science and Engineering 11, 5–7 (2009)

P. Lambrix and G. Kemp (Eds.): DILS 2010, LNBI 6254, pp. 1–2, 2010.
© Springer-Verlag Berlin Heidelberg 2010

9. Freire, J., Koop, D., Santos, E., Silva, C.T.: Provenance for computational tasks: A survey. Computing in Science and Engineering 10(3), 11–21 (2008)
10. Freire, J., Silva, C.: Towards enabling social analysis of scientific data. In: ACM CHI Social Data Analysis Workshop (2008)
11. Freire, J., Silva, C., Callahan, S., Santos, E., Scheidegger, C., Vo, H.: Managing rapidly-evolving scientific workflows. In: Moreau, L., Foster, I. (eds.) IPAW 2006. LNCS, vol. 4145, pp. 10–18. Springer, Heidelberg (2006)
12. Koop, D., Santos, E., Bela Bauer, J.F., Troyer, M., Silva, C.T.: Bridging workflow and data provenance using strong links. In: SSDBM (to appear 2010)
13. Koop, D., Scheidegger, C., Freire, J., Silva, C.T.: The provenance of workflow upgrades. In: IPAW (to appear, 2010)
14. Koop, D., Scheidegger, C.E., Callahan, S.P., Freire, J., Silva, C.T.: Viscomplete: Automating suggestions for visualization pipelines. IEEE Transactions on Visualization and Computer Graphics 14(6), 1691–1698 (2008)
15. Lins, L., Koop, D., Anderson, E.W., Callahan, S.P., Santos, E., Scheidegger, C.E., Freire, J., Silva, C.T.: Examining statistics of workflow evolution provenance: A first study. In: Ludäscher, B., Mamoulis, N. (eds.) SSDBM 2008. LNCS, vol. 5069, pp. 573–579. Springer, Heidelberg (2008)
16. Mesirov, J.P.: Accessible reproducible research. Science 327(5964), 415–416 (2010)
17. Moreau, L., Freire, J., Futrelle, J., McGrath, R.E., Myers, J., Paulson, P.: The open provenance model: An overview. In: Freire, J., Koop, D., Moreau, L. (eds.) IPAW 2008. LNCS, vol. 5272, pp. 323–326. Springer, Heidelberg (2008)
18. Santos, E., Freire, J., Silva, C.: Information sharing in science 2.0: Challenges and opportunities. In: ACM CHI Workshop on The Changing Face of Digital Science: New Practices in Scientific Collaborations (2009)
19. Santos, E., Koop, D., Vo, H.T., Anderson, E.W., Freire, J., Silva, C.T.: Using workflow medleys to streamline exploratory tasks. In: SSDBM, pp. 292–301 (2009)
20. Santos, E., Lins, L., Ahrens, J., Freire, J., Silva, C.T.: Vismashup: Streamlining the creation of custom visualization applications. IEEE Transactions on Visualization and Computer Graphics 15(6), 1539–1546 (2009)
21. Santos, E., Lins, L., Ahrens, J.P., Freire, J., Silva, C.T.: A first study on clustering collections of workflow graphs. In: Freire, J., Koop, D., Moreau, L. (eds.) IPAW 2008. LNCS, vol. 5272, pp. 160–173. Springer, Heidelberg (2008)
22. Scheidegger, C.E., Koop, D., Santos, E., Vo, H.T., Callahan, S.P., Freire, J., Silva, C.T.: Tackling the provenance challenge one layer at a time. Concurrency and Computation: Practice and Experience 20(5), 473–483 (2008)
23. Scheidegger, C.E., Vo, H.T., Koop, D., Freire, J., Silva, C.T.: Querying and creating visualizations by analogy. IEEE Transactions on Visualization and Computer Graphics 13(6), 1560–1567 (2007)
24. Scheidegger, C.E., Vo, H.T., Koop, D., Freire, J., Silva, C.T.: Querying and re-using workflows with vistrails. In: SIGMOD, pp. 1251–1254 (2008)
25. Silva, C., Freire, J., Callahan, S.P.: Provenance for visualizations: Reproducibility and beyond. IEEE Computing in Science & Engineering (2007) (to appear)
26. Silva, C.T., Anderson, E., Santos, E., Freire, J.: Using vistrails and provenance for teaching scientific visualization. In: Proceedings of the Eurographics Education Program (to appear, 2010)
27. Silva, C.T., Freire, J.: Software infrastructure for exploratory visualization and data analysis: past, present, and future. Journal of Physics: Conference Series 25(012100), 15 pages (2008) (SciDAC 2008 Conference)
28. VisTrails, http://www.vistrails.org

High-Performance Systems for in Silico Microscopy Imaging Studies

Fusheng Wang[1], Tahsin Kurc[1], Patrick Widener[1], Tony Pan[1], Jun Kong[1],
Lee Cooper[1], David Gutman[1], Ashish Sharma[1], Sharath Cholleti[1],
Vijay Kumar[2], and Joel Saltz[1]

[1] Center for Comprehensive Informatics
and Department of Biomedical Engineering
Emory University, Atlanta, Georgia, USA
[2] Dept. of Computer Science and Engineering
Ohio State University, Columbus, Ohio, USA

Abstract. High-resolution medical images from advanced instruments provide rich information about morphological and functional characteristics of biological systems. However, most of the information available in biomedical images remains underutilized in research projects. In this paper, we discuss the requirements and design of system support for composing, executing, and exploring in silico experiments involving microscopy images. This framework aims to provide building blocks for large scale, high-performance analytical image exploration systems, through rich metadata models, comprehensive query and data access capabilities, and efficient database and HPC support.

1 Introduction

Technologies for in vitro imaging of biological systems at the microscopic level have advanced significantly in the past decade. Commercial microscopy scanners are now capable of producing high-magnification, high-resolution images from whole slides and tissue microarrays within several minutes. These capabilities reduce dependency on glass slides for expert reviews to assess tissue quality and diagnose disease stage. Moreover, they enable novel *in silico* imaging studies[1] of normal and disease states of biological systems at cellular and subcellular scales. High-resolution image data offers enormous information with which to examine the spatial characteristics and relationships of subcellular structure of specimens under study. A better understanding of those characteristics can lead to better biomarkers or unveil new insights into disease mechanisms.

Software for use of digitized slides in clinical setting is typically characterized by the functionality it provides for a user to browse, view, and manually

---

[1] The term "in silico study" or "in silico experiment" broadly refers to a study or an experiment performed on a computer via analysis, mining, and integration of databases and/or through simulations.

P. Lambrix and G. Kemp (Eds.): DILS 2010, LNBI 6254, pp. 3–18, 2010.

annotate individual slides for tissue quality control and diagnosis. In silico experiments involving image data, on the other hand, have different characteristics and introduce a richer set of data access and processing patterns.

First of all, image data can reach very large volumes. Each image obtained from a whole tissue slide using a state-of-the-art scanner can be tens of gigabytes in size. Large studies may involve thousands of slides obtained from a large cohort of subjects. The sizes of these image datasets can range from terabytes to hundreds of terabytes – it is not too far-fetched to expect that dataset sizes will scale to petabytes, thanks to continued advances in scanning technologies. Such large scale data poses problems in storing, managing, and querying the data.

Second, image data is processed using simple and complex operations and by analysis workflows of various types in in silico experiments. Data processing operations may include filtering, correction of image acquisition artifacts, intensity normalization, registration, segmentation of structures (e.g., nuclei and blood vessels as shown in Figure 1), extraction of features, and classification of segmented structures. These operations can be combined in a variety of ways to form analysis workflows. The sizes of high-resolution images and the complexity of such operations as segmentation and classification may result in long execution times and may require large main memory and powerful computers. Clearly, large scale image analyses are good candidates for execution on parallel and distributed machines.

Third, results from image analyses, whether obtained via manual classification by an expert reviewer or through computer methods, should be expressed in a form that supports efficient synthesis of information. This is necessary to enable sharing and further exploration of results from an in silico experiment, to facilitate comparisons across multiple analyses, and to support rapid development and algorithm evaluations – a large scale study may involve hundreds of methods and analysis workflows. Rich metadata needs to be captured in order to describe analysis results (e.g., nuclear texture, blood vessel characteristics) and the context of the image analyses. With large datasets, researchers have to store, manage, and interact with large volumes of metadata about segmented anatomic structures, markups and features computed for each anatomic object, and semantic information associated with annotations (about cell types, genomic information associated with cells, etc). It is also important to model analytic procedures and pipelines used to carry out segmentation, feature generation, and classification.

Furthermore, comprehensive query support is needed. Researchers would like to query anatomic structures and objects, semantic annotations on objects, and spatio-temporal relationships in order to mine, explore, and correlate the characteristics of specimens under study and integrate with other types of data such as omics and clinical data. A researcher may, for example, want to search for blood vessels by not only shape features like length or thickness but also by their types. In an algorithm evaluation scenario, queries may look for the amount of overlap between objects detected by different algorithms or differences in classification results from an algorithm and a human. A whole slide image may contain millions

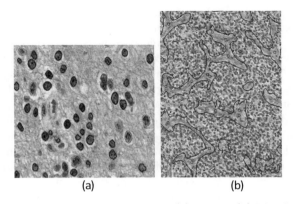

Fig. 1. Examples of image markups: (a) Nuclei; (b) Blood vessels

of anatomic structures, which may have complex shape and texture characteristics, hence there may be millions of annotations associated with the image. A repository of analysis results should be able to support queries on terabytes of image data and hundreds of millions (even billions) of anatomic structures, features, and semantic annotations.

We have presented and discussed solutions to the first and second challenges elsewhere [13,14,12]. In this paper, we propose and discuss a data warehouse framework to support storage, management, and querying of results from in silico image studies. We present an implementation of the core repository infrastructure of the proposed framework. This implementation uses an object-oriented model, called Pathology Analytical Imaging Standard (PAIS) [21], for representation of image analyses. It employs a relational database management system, IBM DB2, for data storage and management.

2   An Example Application Scenario

There are a variety of studies that make use of microscopy imaging, including characterization of the tumor microenvironment and comparative analysis of tissue microarrays. Here, we present a multi-scale integrative research project in cancer research as an example application scenario. We will use this example to illustrate the requirements and design choices to be presented in the following sections.

The In Silico Brain Tumor Research Center (ISBTRC) is a research project funded by the National Cancer Institute as one of the six In Silico Research Centers of Excellence (ISRCE, https://wiki.nci.nih.gov/display/ISCRE). The overarching goal of the ISRCE program is to carry out novel scientific research by analyzing, mining, and integrating publicly available biomedical datasets. The ISBTRC conducts hypothesis-driven translational research on brain tumors. Initially the research will focus on mechanisms for better classification of diffuse gliomas and on the biology of disease progression. This research makes use of complementary genomic, pathology, and radiology brain tumor data from the Cancer Genome Atlas (TCGA), Rembrandt, and Vasari studies.

The ISBTRC is undertaking multiple approaches to study gliomas. One of them involves systematically executing and evaluating in silico experiments to look for relationships between 1) nuclear shape and texture in microscopy images and gene expression profiles defined by molecular clustering analyses and 2) the characteristics of angiogenesis (as detected in microscopy images), gene expression profiles, and neuroimaging features. In the in silico experiments designed for the pathology data, high-resolution images from whole slides are reviewed by expert pathologists as well as analyzed by computer algorithms. The pathologists mark up histological entities of interest on a selected subset of slides, annotate the structures (i.e., assign a classification value), and grade each selected slide. Computer algorithms segment anatomic structures, compute a set of features (ranging from the area and elongation of a nucleus to the bifurcations of blood vessels), and annotate each segmented structure with a semantic classification value (e.g., astrocytoma or oligodendroglioma). The pathologist reviews are used in validation of image analysis methods and to improve the algorithms' segmentation and classification results. Markups and annotations from multiple algorithms also are compared to assess the relative performance of the algorithms.

3   A Framework for in Silico Experimentation with Pathology Images

As we have alluded to in the introduction section, in silico experiments add new data access and processing requirements on top of the basic requirements of viewing and manually annotating individual slides. In a typical study, volumes of image data will be analyzed and mined by computer algorithms to look for morphological patterns that can assist in developing new hypotheses or proving/disproving a hypothesis. Since it may not be feasible to manually examine and classify each slide in a large study, multiple computational methods and workflows may be employed. By comparing and evaluating results from different analyses, a researcher can assign a confidence level to the experiment outcome. The researcher may also design an experiment to rapidly evaluate algorithms in their early stages of development to assess algorithm accuracy and speed. This type of experiment would involve running the algorithms possibly many times against one or more datasets as well as querying, retrieving, and comparing results from other algorithms and previous runs.

We describe at a high-level a software framework to address these types of data access and processing requirements. An illustration of this framework is provided in Figure 2. The framework consists of four main components. We now briefly describe these components.

Analytical Workflow Component. This component implements support for execution of analysis methods and workflows. For large datasets, it should take advantage of parallel and distributed machines and enable data-parallel and

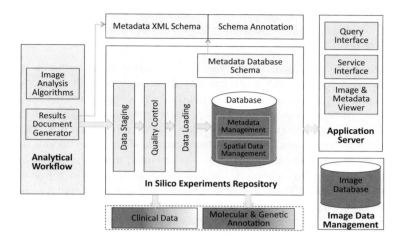

Fig. 2. Analytical microscopy imaging framework

task-parallel implementations of workflows that consist of a network of data processing operations. A subcomponent of this component is the results document generator. Each image analysis application or human annotation application generates the final result data in a format that conforms to the metadata model schema. Provenance information also is encoded in the document; the provenance information could include metadata about algorithm or workflow, analysis parameters, and input and output datasets. For example, in our implementation for the ISBTRC project, results and provenance information are submitted to the results repository as XML documents, conforming to the PAIS metadata model (see Sections 5.1 and 5.2).

Image Data Management. The image database provides the central repository for all microscopy images referenced in a study. To optimize data retrieval speeds for queries on large images and image regions, each image is partitioned into tiles or chunks. These chunks are distributed across multiple disks or storage systems to increase parallel I/O opportunities and are clustered on disks to reduce I/O seek overheads. The images or image tiles are stored in compressed form using a multi-resolution compression scheme in order to reduce storage and I/O costs. Multi-resolution spatial indices, such as R-trees, are employed to reduce the cost of searching the tile set of interest. An implementation of the image data management component is presented in [6].

Application Server. The application server component provides interfaces for query, algorithm invocation, data exchange and sharing, and data viewing. The query interface facilitates a flexible, convenient mechanism to search for and retrieve the data of interest. Additional user defined functions can also be created and run in the database engine, and executed from the query interface in order to provide improved performance. The service interface subcomponent supports Grid and Web Service interfaces for remote access to analyses and for sharing of experiments and methods through well-defined interfaces. Tools and viewers for browsing and viewing image data and analysis results are part of the application server component as well.

In Silico Experiments Repository. This is the central component for management of analysis results, which are generated through computer algorithms or by human experts. The repository is anchored on a data model that consists of generalized data objects, comprehensive data types, and flexible relationships between data objects. In an implementation of this repository, the data model should be designed to capture metadata about in silico experiments, semantic metadata about segmented and classified structures, and provenance information about analyses. The repository instance should be able to allow access to information via a wide range of queries on metadata, spatial structures and relationships, and semantic annotations and relationships drawn from one or more domain ontologies. The in silico experiments repository is the focus of this paper and will be described in greater detail next.

4 Repository for in Silico Microscopy Imaging Experiments

In this section we discuss the requirements and design of repositories for in silico imaging experiments. These repositories first and foremost should enable a research team to efficiently carry out imaging experiments. That is, they should allow for efficient exploration of analysis results. They should also provide support for archiving analyses an investigator wants to save, share, and reference in other studies as well as for agile rapid prototyping and algorithmic exploration.

4.1 Metadata

Rich metadata plays a crucial role in sharing, reusability, and reproducibility of in silico imaging experiments. Metadata should be able to precisely and unambiguously describe an in silico experiment and its components. One of the reasons that information derived from biomedical images is underused can be attributed to lack of efficient and flexible data models to support the modeling, managing, querying and sharing of analysis results and derived data. The Annotation and Image Markup (AIM) model is a caBIG® standard[7] developed to provide standardization for image annotation and markup for radiology images. However, microscopy and pathology images have their unique characteristics.

The immediate challenge is that the metadata model should be efficient to support large volumes of result sets. For instance, one of the ISBTRC experiments involving 213 whole-slide images has segmented and annotated more than 90 million nuclei. In addition, a single XML-based results document, which contained markups for all nuclei and the 23 features associated with each nucleus on a single slide, reached 7GB in size. Another challenge is the complexity of analysis results. The metadata about an in silico experiment can be semantically complex. The metadata model should be able to represent slide related image, markup, feature, and annotation (e.g., classification of anatomic structures) information. This information includes a) context relating to patient data, specimen preparation, special stains, etc, b) human observations involving pathology classification and characteristics, and c) algorithm and human-described segmentations (markups), features, and annotations. Markups can be either geometric shapes or image masks; annotations can be calculations, observations, disease inferences or external annotations. The relationships between data elements can also be complex. For example, additional annotations can be derived from existing annotations. As a result, generic and extensible metadata models are required to support different types of experiments and applications.

The metadata model should also include a semantic description of the computation being carried out. At a minimum, the model should allow a user to express algorithm metadata, parameters, and semantic and concrete identification of input and output datasets. A more advanced model could support ontology-driven semantic descriptions of workflow templates and instances as well as concrete provenance information about an execution of a given workflow. Ontology-driven semantic representations provide a richer system of searching and reasoning about workflows. An example of semantic workflow systems is WINGS [8]. It provides a core ontology for generic components and data types to express workflows. This core ontology can be extended to support data types and data processing components in an application domain [13]. WINGS allows a user to describe an application workflow using semantic properties associated with workflow components and data types at a high-level of abstraction referred to as a workflow template. The workflow template and the semantic properties of the components and data types are expressed using the Web Ontology Language(OWL)[2].

4.2 Query Support

The repository should provide support for metadata based queries (e.g., count nuclei where their grades are less than 3), spatial queries (e.g., find density of nuclei where their grades are between 1 and 3 in selected region of interest), and semantic queries based on reasoning on spatial relationships and/or ontology relationships. The types of queries include: i) retrieval of image data and metadata to obtain data for analytical procedures, ii) queries to compare results generated from different approaches, and validate machine generated results against human observations; iii) spatial queries on assessing relative prevalence of features or classified objects, or assessing spatial coincidence of combinations of features or objects; and iv) queries to support selection of collections of segmented regions, features, objects for further machine learning or content based retrieval applications.

Many of the analytical imaging results are anatomic objects such as lesions, cells, nuclei, blood vessels, etc. Spatial relationships among these objects are often important to understanding the biomedical characteristics of biology systems. Common spatial relationships include containment, intersection or overlap,

---

[2] http://www.w3.org/TR/owl-ref

distance between objects, and adjacency relationships. Besides spatial relationships, another common requirement is to support calculation of coordinate and measurement information (such as computing the area, centroid, perimeter, minimal bounding box) of a markup object. The ISBRTC in silico experiments, for example, generate large volumes of results in the form of segmented regions, markups, annotations, and features. These data elements are stored, managed, and queried for algorithm validation as well as integration with genomics, clinical, and radiology data. Examples of the types of queries are: (1) Find the number of nuclei, which are classified by observer A and whose feature f is within the range of a and b; (2) Which nuclei types preserve nuclei features (distance, shape, etc) between two images; and (3) Which brain tumor nuclei classified by observer A and brain tumor nuclei classified by observer B exhibit spatial overlap in a given region of interest.

Annotations on objects may draw from one or more domain ontologies (e.g., cell ontology to describe different cell types, genome ontology to represent genomic characteristics), creating a semantically rich environment. The repository should allow for querying of data using semantic information. An example query from the ISBTRC studies is "Search for objects with an observation concept (astrocytoma), but also extend it to include all its subclass concepts (gliosarcoma and giant cell glioblastoma)." An important aspect of semantic information systems is the fact that additional assertions (i.e., annotations and classifications) can be inferred from initial assertions (also called explicit assertions) based on the ontology and the semantics of the ontology language. This facilitates a more comprehensive mechanism for exploration of experiment results in the context of domain knowledge. In some cases, it is desirable to extend an ontology with new concepts and properties. That is, a researcher may want to define and add new concepts and classes to the ontology using axioms and rules on existing classes and computable attributes, such as spatial relationships based on distance or relationships between computed features. This would allow incorporation of new knowledge to the system, and might result in new set of inferred annotations (assertions). Combined use of semantic stores/reasoners [11,22,5] and rule engines [10] can offer a repository system capable of evaluating spatial predicates and rules [19,15]. In such a system, the rule engine and the semantic store/inference engine interact to compute inferred assertions based on the ontology in the system, the set of rules, and the initial set of explicit assertions (annotations). Rules that utilize the spatial-temporal relationships might generate new instances of ontological concepts based on the evaluation of the rules. These instances are fed into the semantic inference engine to compute new assertions. The new assertions are input to the rule engine to compute new instances based on rules. This process continues iteratively until no more assertions/instances can be generated.

4.3   High-Performance Computing for Large Data Volumes

In order to scale to large volumes of data, the repository should take advantage of parallel computation power and I/O access. This can be achieved through data

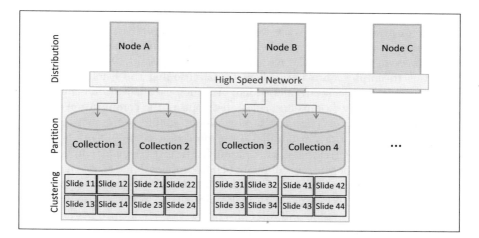

Fig. 3. High performance computing for managing large scale image metadata

distribution and partitioning techniques to take advantage of high performance computing resources (Figure 3) and cluster computing extensions in database management systems,

Data distribution and clustering to reduce I/O costs. Databases can be physically partitioned into multiple physical nodes on cluster based computing infrastructure, which consists of multiple physical servers, where each node has its own CPUs, disk controllers and disks (shared-nothing architecture). Physical database partitions across multiple nodes connected through high speed connections can scale quickly with the power of clusters. Multi-dimensional clustering, on the other hand, provides a method for automatic physically clustering of data along multiple dimensions on more than one key (or dimension) simultaneously. This reduces seek overheads when accessing the data along one or more dimensions. Database logical partitions reside on the same physical node can take advantage of symmetric multiprocessor (SMP) architecture. Having a partitioned database on a single machine with multiple logical nodes is also known as a shared-everything architecture, where the partitions use common memory, CPUs, disk controllers, and disks. Logical partitioned database can then take advantage of the parallelism support for both queries and I/O on a single SMP machine. In addition, *table partitioning* provides another way of dispersing data across multiple storage objects. For example, we can partition data in a table based on slide IDs, or range of dates. This can effectively constrain the search space to boost query performance.

Semantic Query Execution. With very large datasets, semantic query execution and on-the-fly computation of assertions may take too long on a single processor machine to be useful in exploration of datasets. Pre-computation of inferred assertions, also referred to as the materialization process, can reduce the execution of subsequent queries. Materialized assertions can be stored in the system and optimization techniques including indexing can be utilized. However, the process of materialization may take very long for large datasets. Execution strategies leveraging high-performance parallel and distributed machines can reduce execution times and speed up the materialization process [14,19,15]. One possible strategy is to employ data parallelism by partitioning the Euclidean space in which the spatial objects are embedded. Another parallelization strategy is to partition the ontology axioms and rules, distributing the computation of axioms and rules to processors. This partitioning would enable processors to evaluate different axioms and rules in parallel. Inter-processor communication might be necessary to ensure correctness. This parallelization strategy attempts to leverage axiom-level parallelism. It will likely benefit applications where the ontology contains many axioms with few dependencies. A third possible strategy is to combine the first two strategies with task-parallelism. In this strategy, N copies of the semantic store engine and M copies of the rule engine are instantiated on the parallel machine. The system coordinates the exchange of information and the partitioning of workload between the semantic store engine instances and the rule engine instances. The numbers N and M will depend on the cost of the inference execution as well as the partitioning of the workload based on spatial domain and/or ontology axioms.

5   An Implementation of in Silico Imaging Experiments Repository

We have developed an implementation of the in silico experiments repository component (Figure 2) using relational database technology. The database schema is composed of a set of tables based on the Pathology Analytical Imaging Standards (PAIS) model [21]. We describe this implementation in this section.

5.1   PAIS Data Model

The PAIS model is designed to provide an object-oriented, extensible, semantically enabled data model to support pathology analytical imaging and human observations. PAIS provides highly generalized data objects, comprehensive data types, and flexible relationships between data objects. PAIS is also storage and performance efficiency oriented, and supports alternative implementations. Based on an object-oriented design, PAIS is easily extensible. The logical model of PAIS is designed in UML, and consists of 62 classes and interclass associations. The major components (main classes and relationships, not including attributes) are shown in Figure 4. These classes can be categorized as:

- Image reference information – the reference and metadata of the images. These include the ImageReference class with subclasses DICOMImageReference, and MicroscopyImageReference. The later has two subclasses WholeSlideImageReference and TMAImageReference. The Region class specifies which area (e.g., a tile) in the original image is used for the annotation.

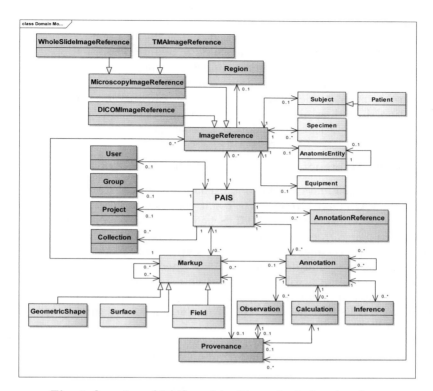

Fig. 4. Overview of PAIS model with a subset of major classes

- Image target information – who, where, and how the images are generated. These include Subject (such as Patient), Specimen, AnatomicEntity and Equipment classes.
- Organizational information – who performs the study and annotation and for what purpose. There are four classes: User, Group, Project and Collection. A collection is a group of items of the same type, gathered for display or study. For example, results from the same algorithm with different parameters on the same image belong to the same collection.
- Markup. Markup delineates a spatial region in the images and represents a set of values derived from the pixels in the images. Markup symbols are associated with one or multiple images, and can be in form of GeometricShape, Surface, or Field. Geometric shapes can be Point, Line, Polyline, Polygon, MultiPoint, MultiLine, MultiPolygon, Rectangle, Circle and Ellipse.
- Annotation. Annotation associates semantic meaning to markup entities through coded or free text terms that provide explanatory or descriptive information. There are three types of annotations: Observation, Calculation, and Inference. Observation holds information about interpretation of a markup or another annotation entity. Observations can be quantified based on different measure scales such as ordinal and nominal scales. Calculation stores

information about the quantitative results from mathematical or computational calculations, represented in CalculationResult, such as Scalar, Array, Histogram, and Matrix. Inference is used to maintain information about disease diagnosis derived by observing imaging studies and/or medical history.
– Algorithm provenance information. The Provenance class, as illustrated in Figure 4, helps to determine the derivation history of a markup or annotation, including algorithm information, parameters, and inputs. Such information is critical for validating approaches and comparing algorithms.

5.2   Results Documents and Data Loading Protocol

To enable convenient data sharing and exchanging, we use XML based representation of the PAIS model to represent result documents. The PAIS logical model is mapped and adjusted into an XML Schema. This schema is used by analysis workflows to generate compatible PAIS XML documents. To reduce the document size for processing, PAIS documents are often generated on partitioned regions such as tiles, and different PAIS document instances from different regions of the same image will share the same document UID. For example, we generate a couple of hundred tile based PAIS XML documents for a single whole slide image. These partitioned PAIS XML documents are further compressed into zipped files.

The XML representation of PAIS facilitates exchange and verification of documents in a standards-based manner. However, for very large result sets, it is not an efficient representation, even with compression of the documents. For exchanging and storing large static data (considered relative to the metadata that will be generated), self-describing structured container technologies (such as HDF5) could provide a more efficient alternative. Such container technologies provide more efficient storage than text-based file formats like XML, while still making available the structure of the data for query purposes. For instance, segmented regions and spatial data structures corresponding to multi-dimensional, multi-resolution data subsets can be stored in HDF5 files.

Data Submission and Staging. PAIS documents generated from image analysis applications can be either submitted to the database server directly by the application on the fly, or grouped for batch submission. PAIS documents are compressed and then submitted to the database server for data staging, where they are stored in a staging table. The database is populated by mapping each XML document into tables. The database internally loads the documents as XML typed column in the database. The temporary XML data enables efficient retrieval for mapping purpose. To map data from XML to the tables, we take advantage of the XMLTABLE function provided by XML databases, which queries the XML column, and generates table like representation of results. These values are then inserted into the tables. To make sure the data loading process works for large transactions, we keep track of the status of each XML document. Initially each document is assigned an "incomplete" status. If a document has been mapped successfully, the status of the document is changed to *complete*. At that point, the XML document can be removed from the database.

5.3 Relational Database Implementation

The database is currently implemented with IBM DB2 Enterprise Edition 9.7.1 with DB2 Spatial Extender, running on PowerEdge T410 (four quadcore CPUs, 16GB memory, and a 15K rpm hard drive) with CentOS 5.5. The database includes (1) *Image target and reference tables* for accessing specimen and image metadata information; (2) *Markup tables* implemented as one spatial table (using the DB2 spatial extensions) that stores geometric shapes in a spatial database, another table that manages association relationship between markups and annotations, and a third table for managing human markups, which are often at a different scale; (3) *Annotation tables* consisting of a table for scalar based calculation results, a table for quantified ordinal scale observations, and another table for quantified nominal scale observations; and (4) *Provenance tables* for managing metadata about algorithms, algorithm parameters, and input datasets.

To support queries on spatial relationships, we model and manage markup objects as spatial objects, supported by spatial databases. In the PAIS data management component, we support the following spatial data types: Point, Line, Polyline, Polygon, Rectangle, Circle, Ellipse, MultiPoint, Multiline, and Multipolygon. The most commonly used spatial type is polygon. These spatial data types are represented as vector graphics based format – SVG[3], so they can be represented as text format for convenient data exchange and visualization. We leverage the spatial extension of DB2 for efficient management and query of spatial information. The spatial table in our implementation is defined as a ST_Polygon spatial data type provided by IBM DB2. We also employ in queries several spatial functions implemented in DB2 such as spatial relationship functions (ST_Contains, ST_Touches, etc) and functions that return information about properties and dimensions of geometries (ST_Area, ST_Centroid, etc). Many of our spatial queries are different from traditional GIS queries. An initial study we have carried out shows that optimizations can be implemented to reduce query execution times. For example, the performance of spatial joins between two algorithms on the same image can be much improved by divide-and-conquer based approach. By dividing a region into four partitions, the cost of spatial overlap queries can be immediately reduced to less than half.

Our current implementation of the application server (see Section 3) offers a SQL interface and a caGrid data service interface [20]. We have developed a caGrid service layer on top of the database to enable data sharing. caGrid is a Grid middleware infrastructure with a service oriented architecture, where researchers can share both their data and analytical resources as grid services, and perform federated queries across distributed databases. We are also building a Google Map like image and metadata viewer, which can quickly zoom into different resolutions of images through identifying and retrieving tiled image portions at specific resolution.

For an initial evaluation of the implementation, we selected 18 slides, and loaded image analysis results from two different algorithm parameter sets and

---

[3] http://www.w3.org/Graphics/SVG

human annotated results. These generate around 18 million markups, and 400 million features. We are able to perform most queries efficiently – the current implementation does not support semantic queries. To support large scale, high performance data management, we plan to use IBM InfoSphere Warehouse Server to manage our data. InfoSphere Warehouse Server uses DB2 with database partitioning features which can effectively support Cluster based and SMP based computing infrastructures.

6   Related Work

Digital microscopy has become an increasingly important biomedical research tool as hardware instruments for rapid capture of high-resolution images from tissue samples have become more widely available. There are several projects that target creation and management of microscopy image databases and processing of microscopy images. The Virtual Microscope system [6] developed by our group provides support for storage, retrieval, and processing of very large microscopy images on high-performance systems. The Virtual Slidebox project [4] at the University of Iowa is a web-based portal of a database of digitized microscopy slides for education. The users can search for virtual slides and view them through the portal. The Open Microscopy Environment project [9] develops a database-driven system for analysis of biological images. The system consists of a relational database that stores image data and metadata. Images in the database can be processed using a series of modular programs. These programs are connected to the database; a module in the processing sequence reads its input data from the database and writes its output back to the database so that the next module in the sequence can work on it. OME provides a data model of common specification for storing details of microscope set-up and image acquisition. OpenCCDB [18,17] is a data model developed to ensure researchers can trace the provenance of data and understand the specimen preparation and imaging conditions that led to the data.

The Allen Reference Atlas (ARA) [1], which is funded by Paul Allen of Microsoft, has a high-resolution anatomical 3-D atlas of the mouse brain. It provides anatomical information for every voxel (at various resolutions of $100 \times 100, 50 \times 50$ down to 25×25) in the 3D coronal atlas made up of 130 coronal mouse slices. The ARA provides a fixed vocabulary of regions names. The Bisque system [16] and associated tools like the Digital Notebook allow a biologist to capture the image experimental data and metadata and store these in a relational database. The eXtensible Imaging Platform (XIP) project is an open source framework for fostering medical imaging algorithm developments [3]. However, this platform is mainly designed and used for radiology image analysis. Additionally, this system lacks a systematic approach for building application workflows, high performance computation and management of image analysis results. DICOM WG 26 is developing a DICOM based standard for storing microscopy images [2], where headers will store metadata such as patient, study and equipment information. Tiles are managed as series and the mapping relationship is represented in an

XML format. However, the metadata is limited and could not be extended for analysis information, and DICOM itself is a storage and data exchange format and not suitable for queries.

7   Conclusion

Availability of an increasing array of high-throughput and high-resolution instruments has given rise to large datasets of omics data – such as genomics, proteomics, metabolomics – and imaging data – such as radiology and microscopy imaging. There are an increasing number of research projects that either primarily focus on in silico experiments or involve them as a significant component of their studies. Microscopy imaging holds tremendous potential for highly detailed in silico examination of morphology of biological systems. In this paper we argue that software for in silico imaging experiments will need to implement more comprehensive support than management, viewing, and annotation of slides. To fully realize the potential of in silico imaging studies, software will be required to support rich, semantic metadata models to represent complex analysis results, databases capable of supporting metadata, spatial, and semantic queries, and high-performance computing techniques for execution of expensive analysis operations and queries.

Acknowledgement. This research is supported in part by PHS Grant UL1RR025008 from the CTSA program, by R24HL085343 from the NHLBI, by Grant Number R01LM009239 from the NLM, by NCI Contract No. N01-CO-12400 and 94995NBS23 and HHSN261200800001E, by NSF CNS 0615155, 79077CBS10, and CNS-0403342, and P20 EB000591 by the BISTI program.

References

1. The allen reference atlas, http://www.brain-map.org, http://mouse.brain-map.org/api/
2. Dicom wg-26, http://medical.nema.org/DICOM/minutes/WG-26/
3. The extensible imaging platform project, https://collab01a.scr.siemens.com/xipwiki/
4. The virtual slidebox, http://www.path.uiowa.edu/virtualslidebox/
5. Broekstra, J., Kampman, A., van Harmelen, F.: Sesame: A generic architecture for storing and querying rdf and rdf schema. In: Horrocks, I., Hendler, J. (eds.) ISWC 2002. LNCS, vol. 2342, pp. 54–68. Springer, Heidelberg (2002)
6. Çatalyürek, Ü.V., Beynon, M.D., Chang, C., Kurç, T.M., Sussman, A., Saltz, J.H.: The virtual microscope. IEEE Transactions on Information Technology in Biomedicine 7(4), 230–248 (2003)
7. Channin, D., Mongkolwat, P., Kleper, V., Sepukar, K., Rubin, D.: The caBIG Annotation and Image Markup Project. Journal of Digital Imaging (2009)
8. Gil, Y., Ratnakar, V., Deelman, E., Mehta, G., Kim, J.: Wings for pegasus: Creating large-scale scientific applications using semantic representations of computational workflows. In: AAAI, pp. 1767–1774. AAAI Press, Menlo Park (2007)

9. Goldberg, I., Allan, C., Burel, J.M., Creager, D., Falconi, A., Hochheiser, H., Johnston, J., Mellen, J., Sorger, P., Swedlow, J.: The open microscopy environment (ome) data model and xml file: Open tools for informatics and quantitative analysis in biological imaging. Genome Biol. 6(R47) (2005)

10. Hill, E.F.: Jess in Action: Java Rule-Based Systems. Manning Publications Co, Greenwich (2003)

11. Kiryakov, A., Ognyanov, D., Manov, D.: Owlim - a pragmatic semantic repository for owl. In: WISE Workshops, pp. 182–192 (2005)

12. Kumar, V.S., Rutt, B., Kurç, T.M., Catalyurek, U.V., Pan, T.C., Chow, S., Lamont, S., Martone, M., Saltz, J.H.: Large-scale biomedical image analysis in grid environments. IEEE Transactions on Information Technology in Biomedicine 12(2), 154–161 (2008)

13. Kumar, V.S., Kurç, T.M., Ratnakar, V., Kim, J., Mehta, G., Vahi, K., Nelson, Y., Sadayappan, P., Deelman, E., Gil, Y., Hall, M., Saltz, J.H.: Parameterized specification, configuration and execution of data-intensive scientific workflows. Cluster Computing (April 2010)

14. Kumar, V.S., Narayanan, S., Kurç, T.M., Kong, J., Gurcan, M.N., Saltz, J.H.: Analysis and semantic querying in large biomedical image datasets. IEEE Computer 41(4), 52–59 (2008)

15. Kurç, T.M., Hastings, S., Kumar, V.S., Langella, S., Sharma, A., Pan, T., Oster, S., Ervin, D., Permar, J., Narayanan, S., Gil, Y., Deelman, E., Hall, M.W., Saltz, J.H.: Hpc and grid computing for integrative biomedical research. IJHPCA 23(3), 252–264 (2009)

16. Kvilekval, K., Fedorov, D., Obara, B., Singh, A., Manjunath, B.S.: Bisque: A platform for bioimage analysis and management. Bioinformatics 26(4), 544–552 (2010)

17. Martone, M.E., Tran, J., Wong, W.W., Sargis, J., Fong, L., Larson, S., Lamont, S.P., Gupta, A., Ellisman, M.H.: The cell centered database project: An update on building community resources for managing and sharing 3d imaging data. Journal of Structural Biology 161(3), 220–231 (2008)

18. Martone, M.E., Zhang, S., Gupta, A., Qian, X., He, H., Price, D.L., Wong, M., Santini, S., Ellisman, M.H.: The cell-centered database: a database for multiscale structural and protein localization data from light and electron microscopy. Neuroinformatics 1(4), 379–395 (2003)

19. Narayanan, S.: Efficient Virtualization of Scientific Data. PhD thesis, Ohio State University, Columbus, OH (2008)

20. Oster, S., Langella, S., Hastings, S.L., Ervin, D.W., Madduri, R., Phillips, J., Kurç, T.M., Siebenlist, F., Covitz, P.A., Shanbhag, K., Foster, I., Saltz, J.H.: cagrid 1.0: An enterprise grid infrastructure for biomedical research. Journal of the American Medical Informatics Association, 138–149 (December 2007)

21. Wang, F., Pan, T., Kurç, T., Sharma, A., Saltz, J.H., Chen, W., Chu, V., Hu, J., Yang, L., Foran, a.D.J.: Unified modeling of image annotation and markup. In: APIII: Advancing Practice, Instruction & Innovation Through Informatics, Pittsburgh, PA (September 2009)

22. Zhou, J., Ma, L., Liu, Q., Zhang, L., Yu, Y., Pan, Y.: Minerva: A scalable owl ontology storage and inference system. In: Mizoguchi, R., Shi, Z.-Z., Giunchiglia, F. (eds.) ASWC 2006. LNCS, vol. 4185, pp. 429–443. Springer, Heidelberg (2006)

Discovering Evolving Regions in Life Science Ontologies

Michael Hartung[1,2], Anika Gross[1,2], Toralf Kirsten[1,3], and Erhard Rahm[1,2]

[1] Interdisciplinary Centre for Bioinformatics, University of Leipzig
[2] Department of Computer Science, University of Leipzig
[3] Institute for Medical Informatics, Statistics and Epidemiology, University of Leipzig
{hartung,tkirsten}@izbi.uni-leipzig.de,
{gross,rahm}@informatik.uni-leipzig.de

Abstract. Ontologies are heavily used in life sciences and evolve continuously to incorporate new or changed insights. Often ontology changes affect only specific parts (regions) of ontologies making it valuable for ontology users and applications to know the heavily changed regions on the one hand and stable regions on the other hand. However, the size and complexity of life science ontologies renders manual approaches to localize changing or stable regions impossible. We therefore propose an approach to automatically discover evolving or stable ontology regions. We evaluate the approach by studying evolving regions in the Gene Ontology and the NCI Thesaurus.

Keywords: ontology evolution, ontology changes, ontology regions.

1 Introduction

Ontologies are heavily used in life sciences, especially to consistently describe or annotate objects of an application domain [1, 14]. For instance, SwissProt [2] and Ensembl [10] are two frequently used data sources in which proteins are annotated (associated) with concepts of the Gene Ontology (GO) [7] to describe their molecular functions as well as their involvement in biological processes. The high importance of ontologies is reflected in their growing number and size. Currently, there are about 70 ontologies available in the Open Biomedical Ontology (OBO) foundry [23]. These ontologies usually underlie a continuous evolution to incorporate the latest requirements and insights of a particular domain [9]. For instance, the GO or the NCI Thesaurus [22] have nearly doubled their size since 2004 [8]. Ontology providers continuously release new versions of changed ontologies. For example, changes for GO are released on a daily basis, and for NCI Thesaurus every month.

As a consequence of this evolution ontology users need to cope with these changes. To determine whether applications or data sources need to be adapted for the newest ontology versions it is valuable to know what parts of an ontology have significantly changed or remained unchanged in a specific period of time. Such information can be utilized in different ways. On the one hand, analysis applications such as functional profiling [3, 21] that used a heavily changed ontology region should be rerun to determine how analysis results are affected by the ontology changes. On the other hand, algorithms may use the information that specific ontology parts remained unchanged

P. Lambrix and G. Kemp (Eds.): DILS 2010, LNBI 6254, pp. 19–34, 2010.

for a more efficient computation since they can reuse previous results. For example, algorithms to match different ontologies [5] can then reuse match results of previous versions for improved efficiency. The information on stable or changing ontology regions is also a good indicator where little or much further development is to be expected. So, unstable ontology regions are a good indicator for ontology developers to participate within a collaborative ontology development. Furthermore, project coordinators may use the information about regions to plan future development steps.

The manual discovery of stable and changing ontology regions is not feasible for large ontologies so that automatic techniques are required. So far only little and preliminary work has been performed in this direction. Previous research in the area of ontology change (see [6] for a survey) focused on ontology versioning [12, 18], the ontology evolution process [15, 24, 25] or the change detection between ontology versions [16, 17, 19, 20]. In our own previous work we quantitatively evaluated evolution of life science ontologies [9]. Furthermore, we designed a web application [8] which allows access to information about changes in life science ontologies. However, to our best knowledge no current work determines the location (region) where changes occurred in an ontology. We therefore make the following contributions in this paper:

- We introduce and define the notion of ontology regions and corresponding measures to classify ontology regions according to their change intensity.
- We propose an algorithm for the discovery of stable and unstable ontology regions. The algorithm is customizable to meet the requirements of different applications. It (1) considers different change types, (2) uses an extensible set of measures for regions and (3) allows region discovery over different time periods. Hence, we can support various application scenarios, e.g., finding small and unstable, or large and stable ontology regions.
- We evaluate the approach for the Gene Ontology and NCI Thesaurus. Results show that in both cases unstable and stable regions exist and hence indicate that the proposed approach is applicable for automatic discovery of evolving regions in large life science ontologies.

The rest of the paper is organized as follows. In Section 2 we present our models for ontologies as well as ontology changes and introduce the notion of ontology regions. Section 3 describes the discovery algorithm. We evaluate the approach in Section 4. We finally conclude and outline possibilities for future work.

2 Preliminaries and Models

We first outline our ontology model including versioning. Next, we describe which kinds of ontology changes are considered and introduce a corresponding change cost model. Finally, we define ontology regions and outline possible measures to quantify the change intensity of regions.

2.1 Ontology Model and Versioning

An ontology $O = (C, R)$ consists of concepts C which are interconnected by relationships in R. Together they form a so-called directed acyclic graph (DAG) representing the structure of O. Special concepts of C called *roots* are the topmost concepts of O, i.e., they have no relationship to any parent concept. If the number of *roots* is greater than one, we introduce a *virtual root* which acts as a single entry point for the ontology. Thus, we can define all *roots* of the ontology as children of the *virtual root*.

A concept $c \in C$ of an ontology is defined by a set of single-valued or multi-valued attributes. The accession number c_{acc} is a special attribute to unambiguously identify ontology concepts. Further typical attributes include the name/label, a definition or synonyms of concepts. Relationships $r \in R$ can be separated into two groups: (1) is_a relationships and (2) other relationships. Is_a relationships usually form the base structure of an ontology, hence we will utilize these relationships to define our ontology regions (see Section 2.3) and make use of them in our discovery algorithm (see Section 3). Other relationships extend the basic is_a structure by more specific relationships, e.g., part_of or has_parts. . The used ontology model represents well existing life science ontologies, in particular the ones in the OBO Foundry [23].

An ontology version $O_v = (C, R, t)$ of version v is a snapshot of an ontology at a specific point in time t. The concepts C and relationships R of O_v are valid until a newer ontology version is released. We assume that versions of an ontology follow a linear versioning scheme, i.e., each ontology version O_i has at most one successor O_{i+1} and one predecessor version O_{i-1}. The first / last ontology versions have no predecessor / successor version, respectively.

2.2 Ontology Changes and Cost Model

The evolution from an old ontology version O_{old} to a newer ontology version O_{new} can be described by a set of ontology changes. We distinguish between the basic change types addition (*add*), deletion (*del*) and update (*upd*) for *concepts*, *relationships* and *attributes* of an ontology:

concept		relationship		attribute		
add	*del*	*add*	*del*	*add*	*del*	*upd*

Particularly, concepts, relationships and attributes can be added or deleted. In case of attributes we further use the update change type for attribute value changes in concepts, e.g., the modification of a concept's name or definition. Note that at the current stage we do not include complex changes such as merge or split of concepts, since these changes are typically composed of basic changes that we already cover. However, complex changes can be included in the future to achieve a more fine-grained and semantically richer distinction between different changes.

To reflect the impact of changes we introduce a *cost model* for ontology changes. Particularly, we assign *change costs* to the different kinds of ontology changes to determine their impact on an ontology. For instance, we can assign higher change costs for *delConcept* compared to *addConcept* to consider a higher change impact for concept deletions vs. concept additions. The individual costs can be assigned to ontology concepts affected by an ontology change. Particularly, we distinguish between

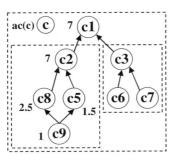

OR	abs_size	rel_size	abs_costs	avg_costs
c1	8	8/8=1	7	7/8=0.875
c2	4	4/8=0.5	7	7/4=1.75
c3	3	3/8=0.375	0	0/3=0

Fig. 1. Sample ontology with regions, aggregated costs (left) and corresponding region measures (right)

two types of costs for an ontology concept: local costs lc and aggregated costs ac. Local costs $lc(c)$ cover the impact of ontology changes that directly affect an ontology concept c, i.e., changes on the concept itself as well as changes on its relationships and attributes. For instance, the addition/deletion of a child concept or an attribute value change have a direct impact. We will later (Section 3.1.1) discuss how local costs are assigned to ontology concepts based on the change type. We further use aggregated costs $ac(c)$ to reflect all changes occurring in the is_a descendants of a concept c. For instance, leaf concept additions/deletions have an indirect impact on corresponding ancestor concepts in the ontology. In Section 3.1.2 we describe how aggregated costs are derived from local costs. The sample (changed) ontology version in Fig. 1 (left) contains aggregated costs (numbers next to a concept) for each concept, e.g., concept $c2$ has aggregated costs of 7 while its sibling $c3$ has no aggregated costs $ac(c3) = 0$.

2.3 Ontology Regions and Measures

An ontology region OR is a subgraph of an ontology consisting of a single root concept rc. A region contains all concepts located in the is_a subgraph of rc, i.e., there exists at least one is_a-path from every concept $c \in OR$ to rc. We will aggregate the concept change costs per ontology region to identify change-intensive or stable regions. Our notion of an ontology region observes that changes often occur in the boundary of an ontology, e.g., addition of leaves or subgraphs to extend the knowledge of a specific topic. Of course an ontology region also covers changes on inner concepts since all intermediate concepts between the root and the leaves are part of the region. In the sample ontology of Fig. 1 (left) several ontology regions are marked. For instance, the region with root concept $c2$ consists of the four concepts $c2$, $c5$, $c8$ and $c9$. The complete ontology with root $c1$ can also be seen as a region.

The change intensity of an ontology region OR and other characteristics can be described by *region measures* incorporating aspects such as the local/aggregated costs or the region size. We will later use these measures in our algorithm for the discovery of specific ontology regions. We define the following exemplary measures for an ontology region *OR*:

- absolute region size *abs_size(OR)*: number of concepts in an ontology region *OR*
- relative region size *rel_size(OR)*: relative size of *OR* compared to the overall size of the ontology *O* defined by *abs_size(OR) / abs_size(O)*
- absolute change costs *abs_costs(OR)*: the absolute costs of *OR* represented by its root's aggregated costs *ac(rc)*
- average change costs *avg_costs(OR)*: the average costs per concept in *OR* defined by *abs_costs(OR) / abs_size(OR)*

Note that these measures are only examples, i.e., we can extend the set of measures depending on application requirements. For instance, one may consider other characteristics such as the depth or the compactness of a region. The example regions *c1*, *c2* and *c3* of the sample ontology in Fig. 1 show different characteristics based on our example measures, as shown in the table on the right side of Fig. 1. For instance, regions *c2* and *c3* have a similar size but differ largely in their change intensity (measures *abs_costs* and *avg_costs*). While region *c3* has not been changed (*avg_costs* of 0), region *c2* exhibits average costs of 1.75. We will now (Section 3) explain how we determine aggregated costs (ac) of concepts in general and for our example ontology of Fig. 1.

3 Ontology Region Discovery

In this section we present the algorithm for discovering evolving ontology regions. We first show how the aggregated costs of concepts are computed for two succeeding ontology versions. We then present the algorithm for the computation of region measures. Finally, we combine both algorithms to discover ontology regions for multiple ontology versions released in a specific period of time.

3.1 Computation of Aggregated Costs for Two Ontology Versions

The algorithm for determining aggregated costs in two succeeding ontology versions takes as input an old ontology version O_{old} and a new ontology version O_{new} as well as change costs σ for ontology changes (see Section 2.2). Note that we use dedicated concept attributes to store local (*lc*) and aggregated costs (*ac*) of concepts, i.e., we internally extend the given ontology versions to capture assigned costs in each concept. The algorithm computeAggregatedCosts consists of four steps as follows:

Algorithm 1: computeAggregatedCosts (ontology versions O_{old}, O_{new}, change costs σ)

ΔO_{old}-O_{new} := diff (O_{old}, O_{new}) computes changes between ontology versions (both directions)

assignLocalCosts (ΔO_{old}-O_{new}, σ, O_{old}, O_{new})

O_{old}:=aggregateCosts (O_{old})

O_{new}:=aggregateCosts (O_{new})

transferCosts (O_{old}, O_{new})

return O_{new}

We first compute the changes between the input versions (diff). Next we assign local costs to affected concepts (assignLocalCosts) to determine the added, deleted and modified ontology elements. Depending on the change type local costs are assigned either to concepts of the older or the newer ontology version. For instance, the deletion of a concept can only be captured in the older version since the concept is not available in the newer one and vice versa for added concepts. Afterwards the local costs are propagated upwards in each ontology version (aggregateCosts) according to the respective ontology structure. This step ensures that costs from deeper ontology parts are aggregated within inner ontology concepts and finally in the ontology root. Since we like to discover regions based on the latest ontology version we need to transfer aggregated costs of older versions to newer ones (transferCosts). The transfer guarantees that costs originated in older ontology versions such as deletes are also reflected in the newest ontology version. Finally, the newer ontology version including the computed aggregated costs is returned. We then can use this ontology version for applying our region measures (see Section 3.2). We also use this enriched version in the iterative algorithm for dealing with more than two ontology versions (see Section 3.3). We will explain the steps of computeAggregatedCosts in more detail in the following sub sections. A simple yet comprehensive example will be used for illustration.

3.1.1 Change Detection and Assignment of Local Costs

Change detection between the two ontology versions O_{old} and O_{new} is based on the comparison of concept accession numbers which are typically used in life science ontologies for unambiguous concept identification. Particularly, we determine ontology changes by comparing elements of O_{old} with those of O_{new}: diff(O_{old}, O_{new}). In this process we distinguish between concept, relationship and attribute changes. Added elements (*add*) are only present in the newer version O_{new} while deleted elements (*del*) only exist in the older version O_{old}. Furthermore, we detect updates (*upd*) on attributes, e.g., when the name of a concept has been changed. Thus, we cover all changes described in Section 2.2. Note that these changes represent the basic change types in ontology evolution and complex changes such as split or merge can be seen as a composition of these.

The example in Fig. 2 shows two ontology versions O_{old} and O_{new} including changes in concepts and relationships (for simplicity we omit attribute changes and focus on is_a relationships). Particularly, from O_{old} to O_{new} two new concepts (*c8*, *c9*) were introduced while one concept (*c4*) was deleted. Corresponding relationships were inserted ((*c8,c2*), (*c9,c5*), (*c9,c8*)) and removed ((*c4,c2*)).

The changes in the diff result and the specified change costs are used to assign local costs (*lc*) to affected concepts in both ontology versions. We assign local costs to concepts using the assignLocalCosts method in the following way. Costs of additions and updates are always captured in the new ontology version. In contrast, costs of deletions are captured in the old ontology version since the affected elements (e.g., a deleted concept) are only present in this version. The costs of concept and attribute changes are directly assigned to the affected concept. For relationship changes the costs are assigned to the source and target concept of a relationship. Note that different costs for the source and target concept can be used.

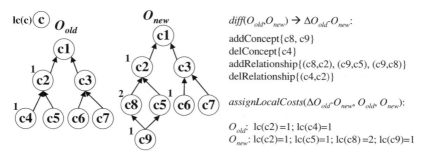

Fig. 2. Diff and assignment of local costs for two ontology versions

In Fig. 2 the numbers next to the concepts refer to the associated local costs for the changes found by diff(O_{old}, O_{new}). For simplicity, we assume uniform change costs of 1 per change. Furthermore, we only assign costs to the target of a changed relationship. In our example the deletion of $c4$ causes the assignment of local costs 1 to $c4$ (*delConcept*) and $c2$ (*delRelationship*) in O_{old}. The insertion of $c8$ and $c9$ (*addConcept*) leads to the assignment of local costs 1 to both concepts. Concept $c8$ receives additional costs 1 caused by the insertion of the ($c9$,$c8$) relationship, thus its overall local costs are 2 (*lc(c8)=2*). Due to the addition of the relationships ($c9$,$c5$) and ($c8$,$c2$) concepts $c2$ and $c5$ of O_{new} are both assigned local cost 1.

3.1.2 Aggregation of Local Costs

We propagate local costs (*lc*) of concepts via is_a paths upwards (in root direction) and hence aggregate costs of subgraphs in corresponding inner ontology concepts (aggregated costs (*ac*) of concepts). The aggregation is applied on the old version as well as the new version with the intention, that the sum of all assigned local costs is equal to the aggregated costs of the root, i.e., the root subsumes all costs assigned to an ontology version.

The aggregation of costs follows one rule. The aggregated costs of a concept is the sum of the aggregated costs of its direct children plus the local costs of itself:

$$ac(c) = \sum_{\text{direct children } c' \text{ of } c} \frac{ac(c')}{|\,parents(c')\,|} + lc(c)$$

If a concept c has more than one parent the *costs* are split into /*parents*/ portions so that *costs*///*parents*/ costs are propagated to each parent. The algorithm aggregateCosts uses an ontology version O_v with associated local costs and propagates them through the ontology using the given structure of O_v:

Algorithm 2: aggregateCosts (ontology version O$_v$)

for all concepts c in O$_v$ **do**

    if local costs lc(c) > 0 **then**

        aggregate (c, O$_v$, lc(c))

    end if

end for

return O$_v$

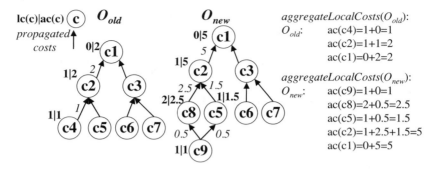

Fig. 3. Aggregation of costs in both ontology versions

Algorithm 3: aggregate (concept c, ontology version O$_v$, change costs σ)

---

aggregated costs ac(c) := ac(c) + σ

parent concepts C$_{parent(c)}$:= getParents(O$_v$, c)

normalized costs σ$_{norm}$:= σ / |C$_{parent(c)}$|

for all concepts c' in C$_{parent(c)}$ **do**

    aggregate(c', O$_v$, σ$_{norm}$)

end for

Fig. 3 shows the aggregation of local costs in our running example for ontology versions O_{old} and O_{new}. Each concept is displayed with its local and aggregated costs ($lc(c)|ac(c)$), paths are annotated with the propagated costs. For instance, in O_{new} $c9$'s aggregated costs are equal to $lc(c9)$ since $c9$ has no children. The relationships $(c9,c5)$ and $(c9,c8)$ are utilized to propagate $c9$'s costs to the corresponding parents. Since two parents exist, $ac(c9)$ is split into two portions of 0.5 which are propagated to $c5$ and $c8$, respectively. Thus, the aggregated costs of $c5$ are composed of $ac(c9)/2$ and $lc(c5)$: $ac(c5) = ac(c9)/2+lc(c5) = 0.5+1 = 1.5$. The same holds for $c8$: $ac(c8) = ac(c9)/2+lc(c8) = 0.5+2 = 2.5$. In the next step $ac(c5)$ and $ac(c8)$ are propagated to $c2$ and aggregated with $lc(c2)$: $ac(c2) = ac(c5)+ac(c8)+lc(c2) = 1.5+2.5+1 = 5$. Having propagated all costs, the aggregated costs of both roots are equal to the sum of all assigned local costs: 2 for O_{old} and 5 for O_{new}, respectively.

3.1.3 Transfer of Aggregated Costs

After separate aggregation of costs in the old and new version the results are now transferred to the newer version. The transfer ensures that change costs of the old version are reflected in the new version as well since we like to discover regions of interest based on the new version. The transferCosts algorithm transfers aggregated costs of concepts from the old version into the new version. In particular, the method

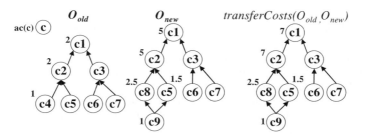

transferCosts(O_{old}, O_{new}):

	ac(c1)	ac(c2)	ac(c3)	ac(c4)	ac(c5)	ac(c6)	ac(c7)	ac(c8)	ac(c9)
O_{old}	2	2	0	1	0	0	0		
O_{new}	5	5	0		1.5	0	0	2.5	1
transfer	7	7	0		1.5	0	0	2.5	1

Fig. 4. Transfer of aggregated costs from old to new ontology version

sums up the aggregated costs of equal concepts in both versions and stores the result in the new version:

Algorithm 4: transferCosts (ontology versions O_old, O_new)

   for all concepts c in O_old **do**

      if c ∈ O_new **then**

         ac(c) ∈ O_new += ac(c) ∈ O_old

      end if

   end for

The transfer of costs for our running example is displayed in Fig. 4. The table below shows how the costs of O_{old} and O_{new} are summed up in O_{new}. Since concepts $c1$, $c2$, $c3$, $c5$, $c6$ and $c7$ are present in the old and new ontology version their aggregated costs of both versions are fused, e.g., after the transfer $c2$'s aggregated costs is 7 (2 from O_{old} and 5 from O_{new}). If a concept is only present in the new version its aggregated costs remain unchanged (e.g., for $c8$ and $c9$ in O_{new}). In contrast, aggregated costs of deleted concepts can not directly be transferred to the new version (e.g., the costs of $c4$). However, the cost aggregation described in Section 3.1.2 ensures that costs of deletions are indirectly transferred. In our case $c2$'s aggregated costs which are transferred to O_{new} contain the costs of $c4$'s deletion. Thus, changes on $c4$ are indirectly reflected in the new version as well.

3.2 Computation of Measures and Discovery of Ontology Regions

To compute the proposed region measures of Section 2.3 we apply an algorithm computeRegionMeasures which uses available information such as aggregated costs or the ontology structure. As an example, in case of the *rel_size* measure we iterate over all ontology concepts and compute the ratio between the region size of each concept and the overall ontology size. $c1$ as the root of our running example exhibits a *rel_size* of 1.0 while $c2$ ($c3$) show a *rel_size* of 0.5 (0.375). As one may notice the sample ontology displayed in Fig. 1 is equal to the result of the transfer of our running example discussed in Section 3.1.3. Hence, the results of our example are equal to the ones presented in Section 2.3.

Based on the results we can discover regions of interest in the new ontology version. Particularly, we define constraints on the results and thus select the regions that satisfy the criteria. Depending on the application different criteria (e.g., relative or absolute size/cost measures) can be considered and combined. For instance, "large stable regions" may be defined with the constraints: *rel_size(OR)*>0.2 and *avg_costs(OR)*=0. In our case region $c3$ is the only region satisfying these constraints. In contrast, one may use *rel_size(OR)*>0.2 and *avg_costs(OR)*>1 to select "large unstable regions", e.g., region $c2$ in our running example. Note that we can eliminate sub-regions of a larger ontology region for a compact result, i.e., only regions satisfying the given constraints and which are not contained in another selected region are returned. For instance, the region covered by $c8$ would also satisfy the constraints of an unstable region (*avg_costs(c8)*>1 and *rel_size(c8)*>0.2). However, $c8$ is contained in region $c2$ and thus we only return $c2$ as an identified region.

3.3 Discovery Algorithm for Multiple Ontology Versions

Based on computeAggregatedCosts and computeRegionMeasures we now define the generalized findRegions algorithm which works on multiple ontology versions released in a specific time period. The idea of the combined algorithm is the following. Having n released ontology versions ($O_1, ..., O_n$) we iterate over all releases and apply computeAggregatedCosts on each pair (O_i, O_{i+1}). Thus, we cover all version changes between succeeding ontology versions and transfer costs from older ontology versions to the latest ontology version O_n where the region discovery is applied (computeRegionMeasures). The algorithm findRegions for n ontology versions looks as follows:

Algorithm 5: findRegions(ontology versions O_1 ... O_n, change costs σ)

---

    for all succeeding ontology versions $O_i - O_{i+1}$ **do**

        O_{i+1} := computeAggregatedCosts(O_i, O_{i+1}, σ)

    end for

    computeRegionMeasures(O_n)

---

4   Evaluation

We evaluated the proposed region discovery algorithm for the well-known Gene Ontology (GO) and the National Cancer Institute Thesaurus (NCIT). After the description of the evaluation setup we first comparatively analyze the overall ontology stability for different periods. In Section 4.3 we analyze the distribution of ontology regions w.r.t. their stability and present how the most (un)stable ontology regions can be discovered. We finally show how the algorithm can be used to track the stability of ontology regions over time.

4.1 Evaluation Setup

The two considered ontologies are heavily used in different projects and underlie continuous changes. GO is widely used for the annotation of proteins w.r.t. Biological Processes (BP), Molecular Functions (MF) and Cellular Components (CC). NCIT maintained at the National Cancer Institute consists of 20 main categories which cover cancer-related topics such as drugs, tissues or anatomical structures. It is utilized in US-wide projects such as the Cancer Biomedical Informatics Grid (caBIG) [4] and its underlying infrastructure caCORE [13].

We integrated available ontology versions between 2004 and 2009 on a monthly basis in a repository [11]. Note that we include at most one version per month, if there is more than one version available we use the first release. The repository allows for the efficient retrieval of versioned ontology information. Thus, we can compare ontology versions of a specified time period to determine the ontology changes in our algorithm. The latest considered GO version of December 2009 consists of 30,304 concepts (GO-BP: 18,108; GO-MF: 9,459; GO-CC: 2,737) while the latest NCIT version of December 2009 contains 77,465 concepts.

For all evaluation studies we apply the following change costs:

concept		relationship		attribute		
add	del	add	del	add	del	upd
1.0	2.0	1.0	2.0	0.5	0.5	0.5

In general concept changes have the biggest impact followed by relationship and attribute changes. Furthermore, we give concept and relationship deletions more impact, attribute changes are weighted equally. In case of relationships we assign half of the costs to the target and the other half to the source concept of a changed relationship. The used values are for illustration only and can be changed to meet specific application characteristics.

4.2 Overall Ontology Stability

We apply our region measures to the root of an ontology for assessing its overall stability. Particularly, we utilize released versions of a specific time period and assess the overall stability by taking the measures *abs_size(root)*, *abs_costs(root)* and *avg_costs(root)* into account. Table 1 lists the overall stability of GO (including its sub ontologies) and NCIT for 2008 and 2009, respectively.

In 2008 GO and NCIT exhibit similar absolute costs (GO: ~24,200; NCIT: ~23,200) but the average change intensity was much higher for GO (*avg_costs* 0.87 for GO vs. 0.32 for NCIT). In 2009, the change intensity increased for NCIT but decreased for GO, but GO still retained an increased change activity (*avg_costs* 0.64 vs. 0.47). Within the GO sub ontologies GO-BP possesses the highest absolute and average costs in both periods. In contrast GO-MF can be seen as the most stable sub ontology of GO ($\leq$0.5 *avg_costs* in 2008 and 2009). Between 2008 and 2009 the average costs decreased especially for GO-MF (from 0.5 to 0.32) underlining the improved stability compared to GO-BP and GO-CC.

Table 1. Overall stability of ontologies in 2008 and 2009

	abs_size(root)		abs_costs(root)		avg_costs(root)	
	2008	2009	2008	2009	2008	2009
GO	27,799	30,304	24,242	19,412	0.87	0.64
– MF	9,205	9,459	4,636	3,002	0.50	0.32
– BP	16,231	18,108	17,594	14,557	1.08	0.80
– CC	2,363	2,737	2,011	1,854	0.85	0.68
NCIT	71,337	77,455	23,165	36,562	0.32	0.47

4.3 Discovery of (un)stable Regions

To discover the most stable and unstable regions of an ontology we analyze the distribution of ontology regions w.r.t. their *avg_costs*. Figure 5 shows such a distribution for GO-BP changes in 2009. We consider ontology regions with a minimum *rel_size* of 0.3% (~ 50 concepts) and group them according to their average costs into intervals of size 0.05. Overall we classified 518 regions in 36 intervals (0.00:0.05 to 1.75:1.80). Most of the regions (~430 regions; ~83%) exhibit average costs between 0 and 0.5, 60 out of which (~12%) have average costs lower than 0.05 and are thus largely stable. In contrast about 53 ontology regions (~10%) show average costs above 0.65.

We can thus determine the most stable and unstable ontology regions by focusing on the two ends of the cost-based distribution. Depending on the application needs we may use either absolute thresholds (e.g., *avg_costs* < 0.01 or *avg_costs* > 0.8) or percentiles of a distribution to classify regions as stable or unstable. For the following analysis, we regard all ontology regions of a certain minimal size below the 5%-percentile as stable and all ontology regions above the 95%-percentile as unstable.

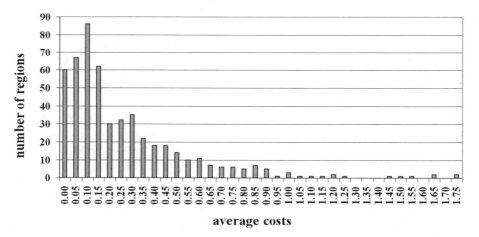

Fig. 5. Distribution of regions w.r.t. average costs for GO-BP in 2009

Table 2. Largest (un)stable ontology regions in 2009

		accession	name	abs_size	rel_size	avg_costs
GO	unstable	GO:0005102	receptor binding	408	4.31%	0.95
		GO:0009653	anatomical structure morphogenesis	583	3.22%	1.22
		GO:0048856	anatomical structure development	566	3.13%	0.91
		GO:0033643	host cell part	77	2.81%	1.90
		GO:0003676	nucleic acid binding	241	2.55%	0.86
		GO:0048646	anatomical structure formation involved in morphogenesis	253	1.40%	0.92
	stable	GO:0031300	intrinsic to organelle membrane	36	1.32%	0.000
		GO:0030054	cell junction	31	1.13%	0.000
		GO:0050865	regulation of cell activation	184	1.02%	0.012
		GO:0075136	response to host	181	1.00%	0.019
		GO:0000151	ubiquitin ligase complex	25	0.91%	0.000
		GO:0016860	intramolecular oxidoreductase activity	71	0.75%	0.000
NCIT	unstable	C28428	Retired Concept	3,264	4.21%	3.49
		C53791	Adverse Event Associated with Infection	1,186	1.53%	2.36
		C45678	Industrial Aid	889	1.15%	1.40
		C74944	Clinical Pathology Procedure	747	0.96%	0.84
		C66892	Natural Product	708	0.91%	1.35
		C53543	Rare Non-Neoplastic Disorder	504	0.65%	1.22
	stable	C64389	Genomic Feature Physical Location	1,026	1.32%	0.000
		C23988	Mouse Neoplasms	886	1.14%	0.000
		C48232	Cancer TNM Finding	742	0.96%	0.000
		C53798	Adverse Event Associated with Surgery & Intra-Operative Injury	707	0.91%	0.000
		C43877	American Indian	555	0.72%	0.000
		C53832	Infection Adverse Event with Unknown Absolute Neutrophil Count	386	0.50%	0.000

Table 2 displays the six largest (un)stable ontology regions of GO and NCIT in 2009. The relative region sizes vary between 0.5% and 5% of the overall ontology size. In GO the relative sizes of the six largest unstable regions are higher than the stable ones. Particularly, the largest stable region in GO exhibits a relative size of 1.32% (GO:0031300) whereas the 6th largest unstable region (GO:0048646) has 1.4% relative size. The largest stable regions regarding absolute size can be found in NCIT consisting of more than 400 concepts. Furthermore, all stable regions of NCIT exhibit no average costs, i.e., in these regions no changes occurred. In contrast, some stable regions of GO show slight average costs, e.g., GO:0050865 or GO:0075136. We further observed that in GO-BP "anatomical structure" topics were highly modified in 2009 (see GO:0009653, GO:0048856 or GO:0048646). Furthermore, in GO-MF the change focus was on special binding functions such as "receptor binding" and "nucleic acid binding". Particularly, "receptor binding" is the largest unstable region of GO (*rel_size*=4.31%). In NCIT "Retired Concept" is the largest unstable region (*rel_size*=4.21%). Note that this ontology region is utilized to collect all ontology concepts that have been retired. Other regions of high interest concern "Drugs and Chemicals" topics such as "Industrial Aid" or "Natural Product".

4.4 Tracking the Stability of Ontology Regions

A sample application of our discovery algorithm is tracking the stability of ontology regions over time. Particularly, we apply our region measures for different time periods to determine the change intensity of different regions over time. We can thus observe certain trends in the evolution of ontologies that are of interest to ontology users.

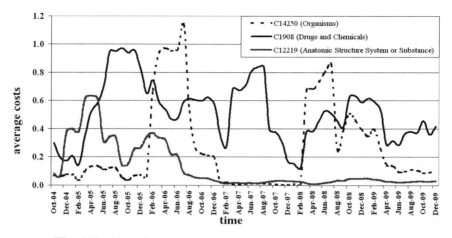

Fig. 6. Tracking of *avg_costs* for sample regions in NCIT (2004-2009)

As an example we applied region tracking on NCIT between 2004 and 2009 for its 20 main categories. The computation uses a sliding window in the following way. We apply our algorithm for a window of size 'half year' (window step: 1 month), i.e., for each window we compute region measures for the selected categories and consider them for a final trend analysis. Hence, we can study variances in the measured results over time, e.g., to find out where and when massive development took place or not.

The chart in Fig. 6 shows the tracking of average costs for three selected main categories of NCIT between 2004 and 2009. We can distinguish different patterns. First, we observe regions, such as "Drugs and Chemicals", that are always unstable, i.e., they experience higher average costs due to frequent modifications. Such regions represent active research fields, and will likely be modified in the near future as well. Furthermore, there are regions such as "Organisms" which exhibit both, periods of high stability mixed with periods of substantial instability. Its instability peaks (Mar 2006-Feb 2007, Mar 2008-Mar 2009) may be caused by new research findings or restructuring decisions by the project consortium which coordinates the ontology development. Finally, there are regions which have become stable over time. For instance, "Anatomic Structure System or Substance" had change activities until the end of 2006, but remained largely stable since 2007. Hence, such a region can be considered as almost finished, i.e., the probability for dramatic changes in the near future is low. This observation especially holds for ontology regions covering accepted / standardized knowledge, e.g., anatomy in the life sciences.

5 Conclusion and Future Work

We introduced the notion of ontology regions and corresponding measures to determine the change intensity or stability of ontology parts. Based on this notion we proposed an algorithm to discover evolving (un)stable regions in life science ontologies by taking ontology changes and the ontology structure into account. The presented algorithm utilizes an adaptable change cost model to reflect the impact of different ontology changes. Our approach can be used in different scenarios, e.g., by ontology users to find out the need to rerun analysis applications or by ontology engineers to notice past and ongoing work in regions of an ontology. We applied our algorithm in a comparative study for two large life science ontologies for different time periods. We observed that the algorithm is able to discover (un)stable ontology regions. The tracking of ontology region stability over time showed different evolution patterns, e.g., ontology regions which are always heavily modified or others that have become stable over the past years.

We see several directions for future work. First, we can consider high-level ontology changes such as merge or split of concepts to achieve a more fine-grained representation of ontology evolution. Second, we plan to integrate the region discovery algorithm into our OnEX system [8]. Finally, we will investigate how algorithms for ontology matching can utilize information about (un)stable regions to determine new ontology mappings in a more efficient way.

Acknowledgments. This work is supported by the German Research Foundation (DFG), grant RA 497/18-1 ("Evolution of Ontologies and Mappings").

References

1. Bodenreider, O., Stevens, R.: Bio-ontologies: current trends and future directions. Briefings in Bioinformatics 7(3), 256–274 (2006)
2. Boutet, E., Lieberherr, D., Tognolli, M.: UniProtKB/Swiss-Prot. Methods in Molecular Biology 406, 89–112 (2007)
3. Boyle, E.I., Weng, S., Gollub, J., et al.: GO:TermFinder - open source software for accessing Gene Ontology information and finding significantly enriched Gene Ontology terms associated with a list of genes. Bioinformatics 20(18), 3710–3715 (2004)
4. caBIG Strategic Planning Workspace: The Cancer Biomedical Informatics Grid (caBIG): infrastructure and applications for a worldwide research community. Studies Health Technology and Informatics 129, 330-334 (2007)
5. Euzenat, J., Shvaiko, P.: Ontology matching. Springer, Heidelberg (2007)
6. Floris, G., Manakanatas, D., Kondylakis, H., et al.: Ontology change: classification and survey. The Knowledge Engineering Review 23(2), 117–152 (2008)
7. The Gene Ontology Consortium: The Gene Ontology project in 2008. Nucleic Acids Research 36(Database issue), D440–D444 (2008)
8. Hartung, M., Kirsten, T., Gross, A., Rahm, E.: OnEX – Exploring changes in life science ontologies. BMC Bioinformatics 10, 250 (2009)
9. Hartung, M., Kirsten, T., Rahm, E.: Analyzing the Evolution of Life Science Ontologies and Mappings. In: Bairoch, A., Cohen-Boulakia, S., Froidevaux, C. (eds.) DILS 2008. LNCS (LNBI), vol. 5109, pp. 11–27. Springer, Heidelberg (2008)
10. Hubbard, T.J., Aken, B.L., Ayling, S., et al.: Ensembl 2009. Nucleic Acids Research 37(Database issue), D690–D697 (2009)
11. Kirsten, T., Hartung, M., Gross, A., Rahm, E.: Efficient Management of Biomedical Ontology Versions. In: Meersman, R., Herrero, P., Dillon, T.S. (eds.) OTM 2009 Workshops. LNCS, vol. 5872, pp. 574–583. Springer, Heidelberg (2009)

12. Klein, M., Fensel, D.: Ontology versioning on the Semantic Web. In: Proceedings of the International Semantic Web Working Symposium (SWWS), pp. 75–91 (2001)
13. Komatsoulis, G.A., Warzel, D.B., Hartel, F.W., et al.: caCORE version 3: Implementation of a model driven, service-oriented architecture for semantic interoperability. Journal of Biomedical Informatics 41(1), 106–123 (2008)
14. Lambrix, P., Tan, H., Jakoniene, V., Strömbäck, L.: Biological Ontologies. In: Semantic Web: Revolutionizing Knowledge Discovery in the Life Sciences, pp. 85–99 (2007)
15. Noy, N., Chugh, A., Liu, W., et al.: A Framework for Ontology Evolution in Collaborative Environments. In: Cruz, I., Decker, S., Allemang, D., Preist, C., Schwabe, D., Mika, P., Uschold, M., Aroyo, L.M. (eds.) ISWC 2006. LNCS, vol. 4273, pp. 544–558. Springer, Heidelberg (2006)
16. Noy, N., Klein, M.: Ontology evolution: Not the same as schema evolution. Knowledge and Information Systems 6(4), 428–440 (2004)
17. Noy, N., Musen, M.: Promptdiff: a fixed-point algorithm for comparing ontology versions. In: Proc. 18th Intl. Conference on Artificial Intelligence, pp. 744–750 (2002)
18. Noy, N., Musen, M.: Ontology versioning in an ontology management framework. IEEE Intelligent Systems 19(4), 6–13 (2004)
19. Papavassiliou, V., Flouris, G., Fundulaki, I., et al.: On Detecting High-Level Changes in RDF/S KBs. In: Bernstein, A., Karger, D.R., Heath, T., Feigenbaum, L., Maynard, D., Motta, E., Thirunarayan, K. (eds.) ISWC 2009. LNCS, vol. 5823, pp. 473–488. Springer, Heidelberg (2009)
20. Plessers, P., De Troyer, O.: Ontology Change Detection Using a Version Log. In: Gil, Y., Motta, E., Benjamins, V.R., Musen, M.A. (eds.) ISWC 2005. LNCS, vol. 3729, pp. 578–592. Springer, Heidelberg (2005)
21. Prüfer, K., Muetzel, B., Do, H.H., et al.: FUNC: a package for detecting significant associations between gene sets and ontological annotations. BMC Bioinformatics 8, 41 (2007)
22. Sioutos, N., de Coronado, S., Haber, M.W., et al.: NCI Thesaurus: A semantic model integrating cancer-related clinical and molecular information. Journal of Biomedical Informatics 40(1), 30–43 (2007)
23. Smith, B., Ashburner, M., Rosse, C., et al.: The OBO Foundry: coordinated evolution of ontologies to support biomedical data integration. Nature Biotechnology 25(11), 1251–1255 (2007)
24. Stojanovic, L., Maedche, A., Motik, B., et al.: User-driven ontology evolution management. In: Gómez-Pérez, A., Benjamins, V.R. (eds.) EKAW 2002. LNCS (LNAI), vol. 2473, pp. 285–300. Springer, Heidelberg (2002)
25. Stojanovic, L., Motik, B.: Ontology evolution within ontology editors. In: Proceedings of the International Workshop on Evaluation of Ontology-based Tools, pp. 53–62 (2002)

On Matching Large Life Science Ontologies in Parallel

Anika Gross[1,2], Michael Hartung[1,2], Toralf Kirsten[2,3], and Erhard Rahm[1,2]

[1] Department of Computer Science, University of Leipzig
[2] Interdisciplinary Centre for Bioinformatics, University of Leipzig
[3] Institute for Medical Informatics, Statistics and Epidemiology, University of Leipzig
{gross,hartung,rahm}@informatik.uni-leipzig.de,
tkirsten@izbi.uni-leipzig.de

Abstract. Matching life science ontologies to determine ontology mappings has recently become an active field of research. The large size of existing ontologies and the application of complex match strategies for obtaining high quality mappings makes ontology matching a resource- and time-intensive process. To improve performance we investigate different approaches for parallel matching on multiple compute nodes. In particular, we consider inter-matcher and intra-matcher parallelism as well as the parallel execution of element- and structure-level matching. We implemented a distributed infrastructure for parallel ontology matching and evaluate different approaches for parallel matching of large life science ontologies in the field of anatomy and molecular biology.

Keywords: ontology matching, matching performance, parallel matching.

1 Introduction

Ontologies and their applications have become increasingly important especially in the life sciences [19, 5]. Typically they are utilized to semantically annotate molecular-biological objects such as proteins or pathways. For instance, the popular Gene Ontology (GO) [10] is the primary ontology for annotating proteins with information on the functions and processes they are involved in. Other life science ontologies, e.g., in the Open Biomedical Ontologies Foundry (OBO) [31] contain information about anatomical structures for different species (e.g., human, mouse, fly) or diseases. The increasing number and availability of different life science ontologies enables new types of analysis, experiments and applications.

Recently, the development and maintenance of ontology mappings interconnecting different (multiple) related ontologies have gained importance, e.g., to integrate heterogeneous information sources (e.g., [15]), to merge ontologies [18], or to support analysis such as the comparison of expression patterns [2]. Since the manual creation of such ontology mappings is time-consuming or even infeasible their semi-automatic generation called *ontology matching* [24, 9] has become an active research field especially for life science ontologies (e.g., [22, 4, 17, 26]).

Effective ontology matching, i.e. the computation of high quality mappings, typically entails the combined execution of several matchers to determine the similarity between ontology elements based on metadata or instance data (see [24, 9]). For large

P. Lambrix and G. Kemp (Eds.): DILS 2010, LNBI 6254, pp. 35–49, 2010.
© Springer-Verlag Berlin Heidelberg 2010

ontologies these matchers are often very time-consuming and memory-intensive. This is because metadata-based matchers, e.g., comparing the names of ontology concepts, typically evaluate the Cartesian product of all element pairs leading to a quadratic complexity w.r.t. ontology size. The performance requirements are further multiplied by the number of different matchers or when applying ontology matching on multiple ontology versions [12, 32]. Ontology matching is also memory-intensive for large ontologies because matching is typically performed on memory representations (graph structures) of the ontologies and requires the maintenance of several similarity values for every element pair from the Cartesian product.

The results of previous OAEI contests [20] on matching anatomical ontologies have shown that systems need execution times of up to several hours. This is despite the fact that the considered ontologies are only of medium size of around 3,000 elements (Mouse Anatomy Ontology [13] with ~2,800 elements was matched against the anatomy part of the NCI Thesaurus [30] with ~3,300 concepts). The Cartesian product thus has about $9 \cdot 10^6$ element pairs to be evaluated. Larger ontologies lead to even higher resource requirements. For instance, matching the two sub ontologies Molecular Functions and Biological Processes of GO with 10,000 and 20,000 ontology concepts results in approx. $2 \cdot 10^8$ pairs to compare, i.e., 22 times more than in the OAEI match problem. The memory requirements just for the similarity values are in the order of several GB.

These examples illustrate that it is valuable to have a match system providing high-performance ontology matching especially for interactive (online) applications where fast response times are required or when multiple match configurations have to be evaluated. While improving ontology matching performance has received some attention recently (see Related Work section), to the best of our knowledge the parallel execution of ontology matching on multiple compute nodes has not been studied so far. However, the broad availability of multi-core systems and multiple computing machines makes parallel ontology matching very attractive. Partitioning a large match problem into smaller parallel match tasks also helps to reduce the memory requirements per task. We therefore study strategies for parallel ontology matching and make the following contributions in this paper:

- We propose different strategies for parallel ontology matching, in particular *inter-* and *intra-matcher parallelization*. While the former approach executes independent matchers in parallel, the latter performs an internal parallelization of matchers based on a partitioning of the ontologies to be matched. Both strategies can be combined for additional performance improvements.
- We show how different kinds of matchers (element-level, structure-level, instance-based matchers) can be parallelized.
- We implemented a distributed infrastructure for parallel ontology matching and evaluate different approaches for parallel matching of large life science ontologies in the field of anatomy and molecular biology. The results show the effectiveness and scalability for single matchers and complete match strategies.

The rest of the paper is organized as follows. In Section 2 we introduce our ontology model and provide background information on ontology matching. Section 3 discusses inter- and intra-matcher parallelization and outlines how different matchers can be executed in parallel. The infrastructure for parallel ontology matching is presented in Section 4. We evaluate our approaches in Section 5 and discuss related work in Section 6. Finally, we summarize and outline possibilities for future work.

2 Preliminaries

We first introduce our ontology model. We then discuss the ontology matching problem and common match approaches.

2.1 Ontology Model

An ontology $O = (C, R)$ consists of concepts C which are interconnected by directed relationships in R. A special concept called *root* has no relationships to any parent. The directed relationships can be of different type. The most common relationship type in ontologies is '*is_a*' describing an inheritance between two concepts. Furthermore the '*part_of*' relationship type is used to model part-whole relationships between concepts. Life science ontologies use further semantic relationship types, e.g. '*regulates*'. We allow several parents and therefore several root paths per concept. The structural information (context) of concepts is used by structure-based match approaches to determine the concept similarity.

Furthermore, a concept $c \epsilon C$ of an ontology is defined by a set of single- or multi-valued attributes. For instance, the concept name is a single-valued attribute that is frequently used for ontology matching. Some ontologies (e.g., GO) support multi-valued synonym attributes containing alternate names for a concept. Usually there is also an identification attribute or accession number c_{acc}. These concept identifiers are used for annotating biological objects (proteins, genes, etc.) [11] and can be useful for instance-based ontology matching.

2.2 Ontology Matching

Ontology matching is the process of determining a set of semantic correspondences (ontology mapping) between concepts of two related ontologies O_1 and O_2. The correspondences are determined by matcher algorithms determining the similarity $sim(c_1,c_2) \epsilon [0...1]$ between concepts $c_1 \epsilon O_1$ and $c_2 \epsilon O_2$. Matchers can roughly be classified into *metadata- or schema-based* and *instance-based* approaches [24]. Metadata-based matchers do not utilize instance data but focus on ontology information and optionally some background information such as dictionaries. Metadata-based matchers can be further classified into *element-level* and *structure-level* matchers. Element-level matchers utilize information from concept attributes, such as determining the similarity of concept names and synonyms, e.g., based on some string similarity such as ExactMatch, n-Gram or EditDistance. Element-level matchers are almost always used and combined with other approaches. *Structure-level matchers* consider the ontology structure for matching, e.g., to determine the context similarity of concepts. Typical matchers evaluate the children, leaves, siblings and ancestors of concepts. In contrast, *instance-based matchers* do not depend on the ontology metadata but utilize existing associations between ontology concepts and instances and consider two concepts as similar if they share similar instances. One way to determine instance similarity is to measure the degree of instance overlap between concepts, e.g., based on a Dice or Jaccard measure. The complexity of matchers is usually quadratic by comparing all concepts of the first ontology with all concepts of the second ontology (evaluation of the Cartesian product).

A single matcher is typically not sufficient for high match quality so that one has to combine several matchers within a so-called match strategy or workflow. Match prototypes such as COMA++ therefore provide many matchers and support their flexible combination [1, 6, 7]. The matchers may be sequentially executed so that the results of a first matcher are refined by the following matchers. Alternatively, the matchers are independently executed and combined. Match workflows may use different methods to combine match results of individual matchers, e.g., by performing a union or intersection or by aggregating individual similarity values. The final match result is typically restricted to correspondences for which the similarity values exceed a predetermined threshold. In the next section, we discuss how such match strategies as well as single matchers can be parallelized.

3   Parallelization Strategies

In this section, we discuss possibilities of parallelizing ontology matching workflows consisting of several matchers that are either sequentially or independently executed. We assume that a computing environment of multiple locally interconnected multicore computing nodes is available for matching.

A straight-forward approach to parallel ontology matching is *inter-matcher parallelism*, i.e., to process independently executable matchers in parallel on different cores or computing nodes. In addition, we want to support *intra-matcher parallelism*, i.e., the internal parallelization of individual matchers. Furthermore, we can combine both kinds of parallelism. In the following, we discuss these parallelization strategies in more detail. For intra-matcher parallelism (Section 3.2) we focus on the parallel similarity evaluation of the Cartesian product of concept pairs according to a partitioning of the input ontologies. In particular we will describe how we can parallelize element-level, structure-level and instance-based matchers.

3.1   Inter-matcher Parallelization

Inter-matcher parallelization enables the parallel execution of independently executable matchers to utilize multiple processors for faster match processing. The example match workflow in Figure 1a utilizes inter-matcher parallelization for n matchers $(M_1, \ldots, M_n)$. The match results can be combined by different aggregation and selection strategies to achieve the final result. Ideally, the inter-matcher parallelization improves the execution time by a factor n if the matchers are of similar complexity. This kind of parallelism is easy to support and can utilize multiple cores of a single computing node or multiple nodes. However, inter-matcher parallelization is limited by the number of independently executable matchers. Furthermore, matchers of different complexity may have largely different execution times limiting the achievable speedup (the slowest matcher determines overall execution time). Moreover, the memory requirements for matching are not reduced since matchers evaluate the complete ontologies.

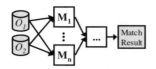

Fig. 1a. Inter-matcher parallelization

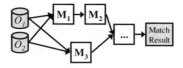

Fig. 1b. Combination of inter-matcher parallelization and sequential matching

The degree of parallelism is also limited for sequential matcher execution (e.g., if a structure-level matcher depends on a previously executed element-level matcher) or when the number of available processors is smaller than the number of independently executable matchers. As illustrated in Figure 1b, in such cases inter-matcher parallelism can be applied for a subset of matchers. The shown example assumes that only two cores can be utilized and that the most complex matcher M_3 is assigned to one core while M_1 and M_2 are executed sequentially on the other core.

3.2 Intra-matcher Parallelization

Intra-matcher parallelization deals with the internal decomposition of individual matchers or matcher parts (e.g., tokenization of concept names) into several match tasks that can be executed in parallel. We focus on a general approach to support intra-matcher parallelism based on partitioning the input data (the ontologies). Such a partitioning is very flexible and scalable and can be used to generate many match tasks of limited complexity. Furthermore, intra-matcher parallelism can be applied for sequential as well as independently executable matchers, i.e., it can also be combined with inter-matcher parallelism.

Figure 2 illustrates intra-matcher parallelization for n matchers that are executed in parallel (i.e., in combination with inter-matcher parallelism). For each matcher the input ontologies are first partitioned followed by the generation of multiple match tasks $M_{i1}, \ldots, M_{ik}$ ($i = 1, \ldots, n$). These match tasks are executed in parallel, the union of the match task results gives the complete match result. In the example, all match tasks of the n matchers can be concurrently executed on the available compute nodes to achieve a maximal reduction of the execution time. Note that the match tasks only match partitions of the two ontologies and have thus reduced memory and processing requirements compared to a complete matcher. Hence, intra-matcher parallelization is especially promising for matching large ontologies.

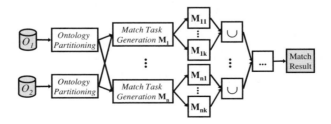

Fig. 2. Intra-matcher parallelization

Before we discuss how we can parallelize element-level, structure-level and instance-based matchers we first outline our approach for *ontology partitioning*. In this initial study of parallel ontology matching we focus on a simple but yet flexible size-based approach that enables the parallel matching of the Cartesian product of the concepts from the two input ontologies O_1 and O_2. To generate match tasks of similar complexity we partition both ontologies into partitions of equal size (number of concepts); the partition size is a parameter that can be chosen according to the size of input ontologies and the complexity of the utilized matcher. Each task matches one O_1 partition with one O_2 partition so that we generate $p_1 \cdot p_2$ match tasks for p_1 (p_2) equally sized partitions of O_1 (O_2). For instance, if we partition two ontologies of 10,000 concepts into 10 partitions each, we generate 10·10=100 match tasks. As we will discuss in Section 4, generated match tasks are managed in job queues from where they are scheduled for parallel execution.

This size-based ontology partitioning has significant advantages besides its simplicity: (1) it is scalable to large ontologies by choosing manageable partition sizes and thus enables unproblematic processing and reduced memory requirements per match task, (2) it supports good load balancing because of equally sized partitions and match tasks, (3) it helps optimizing performance without sacrificing match quality since the full Cartesian product is evaluated, and (4) it can be utilized for element-level, structure-level and instance-based matchers as we will discuss in the following.

3.2.1 Parallelization of Element-Level Matchers

To parallelize element-level matching approaches based on the introduced size-based partitioning is relatively easy. This is because element-level matchers compare ontology concepts with each other by utilizing metadata from the concepts themselves, i.e., their attribute values such as the name or synonyms. By partitioning the ontologies into subsets of concepts we retain the information needed for matching the concepts. Hence, element-level matchers can easily be applied to ontology partitions.

Figure 3 shows a running example for matching two ontology parts $c_1, ..., c_3 \in O_1$ and $d_1, ..., d_5 \in O_2$. As shown, concept c_1 has two children c_2 and c_3. The concept d_3 of O_2 is assumed to have two parent concepts d_1, d_4 (multiple inheritance). Some concepts have associated instances that will be considered later for instance-based matching. We assume that the concepts should be matched with each other by a

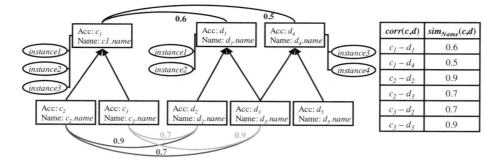

Fig. 3. Element-level matching on *Name* attribute

string-based name matcher. The name matcher evaluates the string similarity (e.g., TriGram) for all ($3 \cdot 5 = 15$) concept pairs. The result set (shown on the right of Figure 3) contains six correspondences with similarities ranging from 0.5 to 0.9; all other concept pairs are assumed to have similarity 0, i.e., they do not match.

3.2.2 Parallelization of Structure-Level Matchers

Structure-level matchers are more difficult to parallelize than element-level matchers since they utilize information from the structural context or neighborhood of concepts (e.g., children, parents, siblings) or even the whole ontology. Hence, an ontology partition consisting of a certain number of concepts does generally not provide all information needed for structure matching. Even more difficult is the parallelization of iterative structural matchers such as Similarity Flooding [21] that start with initial element-level similarities and iteratively propagate these along the concept relationships across the whole ontologies. For such matchers parallelization is inherently difficult and has likely to be restricted to the initial element-level matching.

We therefore focus on structural matchers that utilize information from a restricted neighborhood (local context) of concepts. To limit the resource and memory requirements we do not want the match tasks to work on the whole ontologies but to restrict them to input partitions of restricted size similar to parallel element-level matching. This can be achieved by extending the concept-level information, within special multi-valued *context attributes*, by information from the local context that is needed for structure-level matching. The values for these context attributes, e.g., *Child*, *Parents*, *NamePath,* are determined in a preprocessing step by traversing the input ontologies once (linear effort) to collect the necessary context information about children, parents, etc. Concepts with these additional context attributes can then be partitioned as for element-level matching. Each match task performs structure matching for a pair of partitions utilizing information from the context attributes.

Figure 4 illustrates the context attribute approach for a Children matcher for our running example of Figure 3. The matcher determines the similarity between two concepts by calculating the average element (e.g., name) similarity between their children, i.e. it takes the sum of the name similarities between any two children and divides by the total number of child pairs. Note that this is only one possibility to compute the children similarity, used for illustration. For the example of Figure 4, we obtain that c_1 is more similar to d_1 than to d_4 as c_1 and d_2 share more similar children (using the similarity values of Figure 3). To execute this matcher we use a multi-valued *Child* context attribute for each (non-leaf) concept and populate it during the preprocessing step, in our case with the name values of child concepts. A child match task matches each concept c of an O_1 partition with each concepts d of an O_2 partition by merely comparing all *Child*-attributes of c with all *Child*-attributes of concept d w.r.t. their string (name) similarity and dividing it by number of possible child pairs: $sim_{Children}(c,d) = \sum_{i,j} sim_{Name}(c.child_i, d.child_j) / (|c.child| \cdot |d.child|)$.

The context attribute approach can similarly be applied for other local context matchers such as Parents, Siblings or NamePath. For instance, to realize the NamePath matcher we determine a concept's predecessors in a root path and store their concatenated names in a multi-valued *NamePath* context attribute during preprocessing. Matching is then similar to name element-matching but uses the *NamePath*

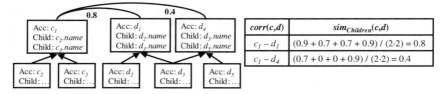

Fig. 4. Attribute-based child matching

attribute and its structural information about the names of the predecessor concepts. In previous evaluations [7], NamePath was shown to be one of the most effective single matchers so that it is valuable to have a parallel implementation of it.

3.2.3 Parallelization of Instance-Based Matchers

Finally, we discuss how instance-based matching approaches can be parallelized. One common approach evaluates the instances associated to ontology concepts and considers two concepts as similar if they largely share similar instances [17]. Since instances are directly associated to concepts, we can determine the concept similarity using concept-specific information. This allows us to apply a similar parallelization strategy as for local-context structure matching and element-level matching.

As illustrated in Figure 5 instances are mapped to a multi-valued attribute *Instance* during preprocessing. For example, *Instance* may contain the accessions of biological objects associated to a GO concept. Size-based partitioning is applied to the input ontologies and the associated instances. A Dice-based measuring of the instance overlap similarity [17] would count the common *Instance* attribute values N_{cd} of two concepts $c \in O_1$, $d \in O_2$, and compute the similarity $sim_{Dice}(c,d) = 2 \cdot N_{cd} / (N_c + N_d)$ where N_c (N_d) is the number of instances of concept c (d). In our example (Figure 5) the match result contains two correspondences with a higher similarity for concepts c_1-d_1 sharing more common instances than c_1-d_4.

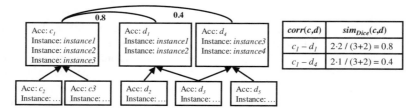

Fig. 5. Attribute-based instance-based matching

4 Infrastructure for Parallel Ontology Matching

To execute ontology matching workflows in parallel we have implemented a distributed and service-based infrastructure illustrated in Figure 6. It consists of several services including a central *workflow service*, a *data service*, and multiple *match services* that are implemented in Java. These services run on different loosely coupled servers or workstations. While the workflow service coordinates the execution of the complete match workflow, match services compute the ontology mapping for two ontologies or ontology partitions. The data service manages all ontology and instance data forming the input of a match workflow and stores the final ontology mapping as result. The data service implements the repository schema proposed in [16] to efficiently store ontology and mapping versions.

Match applications (e.g., matching tools such as COMA++ [1]) use the workflow service to centrally access the match infrastructure. We assume that these applications configure a concrete match workflow, i.e., they specify the ontologies and instance data (or versions of both) as input data as well as utilized matchers and steps to pre-process the matcher input and to post-process match results (e.g., ontology partitioning and mapping manipulations including union and majority as well as filtering). Within this specification the match and manipulation steps are interconnected such that the workflow defines which matchers can be executed in parallel (inter-matcher parallelization) or in sequential order. The workflow service takes this configuration as input and processes the specified match workflow.

The workflow service performs ontology preprocessing if necessary, in particular for determining the values of context attributes for structure-level matching (see Section 3.2). The workflow service executes the matchers in the workflow in the specified order for sequential matchers or in parallel. For this purpose, it maintains a job queue for each matcher. For intra-matcher parallelism the workflow service generates all match tasks and stores them in the matcher-specific job queue. Without intra-matcher parallelization the job queues consist of only a single match(er) job. The workflow service sends the queued match jobs to available match services as long as there are unprocessed jobs available. The match services execute the jobs and send

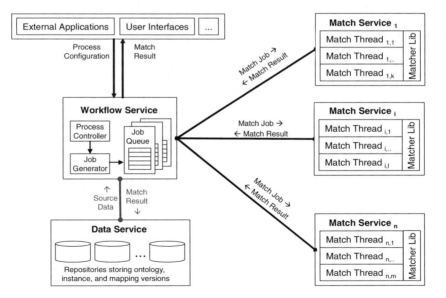

Fig. 6. Distributed infrastructure for matching ontologies in parallel

their results back to the workflow service which unifies the partial match results. For efficiency reasons, the match jobs are restricted by a similarity threshold so that they only return concept correspondences exceeding the minimal similarity.

The match services run on dedicated nodes to fully exploit their compute power. Each match service contains several concurrently working *match threads* executing the match jobs (one thread per job at a time). The number of match threads per service can vary according to the number of available cores on the node. Hence, the infrastructure can cope with heterogeneously configured computing environments, i.e., servers and workstations with different number of cores and speed can be used for the proposed infrastructure. The match threads obtain their input data (ontology partitions) from the match job and execute the specified matcher implementation from a comprehensive *matcher library*.

5 Evaluation

We used the ontology matching infrastructure to evaluate the proposed parallelization strategies. We first describe the evaluation setup, in particular the considered ontologies and matchers. In Section 5.2 we show results for parallel matching on a single multi-core node. We then analyze the scalability of parallel ontology matching on multiple compute nodes.

5.1 Evaluation Setup

We use up to four nodes for running match services each consisting of four cores, i.e., we utilize up to 16 cores. Each node has an Intel(R) Xeon(R) W3520 4x2.66GHz CPU, 4GB memory and runs a 64-bit Debian GNU/Linux OS with a 64-bit JVM. We use 3GB main memory (heap size) per node. The workflow and data services run on additional nodes.

In our experiments we consider a medium-scale as well as a large-scale match problem. For the medium-scale problem we match the AdultMouseAnatomy (MA) (2,737 concepts) with the anatomical part of the NCI Thesaurus (NCIT) (3,289 concepts) as in the OAEI 2009 contest. The large-scale match problem computes an ontology mapping between the two GO sub ontologies Molecular Functions (MF) and Biological Processes (BP) consisting of 9,395 and 17,104 concepts, respectively (versions of June 2009). For intra-matcher parallelism we use different partition sizes for the two match problems. For the medium-scale problem we set the maximum partition size to 500 concepts resulting in 6 (7) partitions for MA (NCIT) and thus 42 match tasks. For the large-scale match problem the max. partition size is set to 1,500. Hence, MF (BP) is split into 7 (12) partitions which results in 84 match tasks.

In this first evaluation analysis we focus on element-level and structure-level matchers. We applied three different single matchers namely NameSynonym (NS), Children (CH) and NamePath (NP). NS determines the maximal TriGram similarity for the name (label) and multi-valued synonym attribute values between concepts. CH and NP use TriGram similarity on the context attributes *Child* and *NamePath*, respectively (see Section 3.2.2). NamePath is restricted to at most three ancestor levels

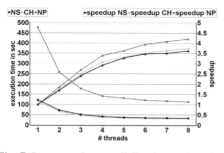

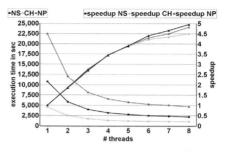

Fig. 7. Intra-matcher parallelization on 1 node: medium-scale problem

Fig. 8. Intra-matcher parallelization on 1 node: large-scale problem

including '*is_a*' and '*part_of*' paths. These matchers are evaluated individually as well as within combined match strategies. In this study we focus on evaluating the efficiency (execution times) and not the matching effectiveness (e.g., precision, recall). This is because our parallel match approaches only target efficiency but do not affect quality since we always evaluate the Cartesian product, e.g. when using size-based partitioning (as described in Section 3.2).

5.2 Individual Matcher Parallelization on a Multi-core Node

We first analyze intra-matcher parallelization of individual matchers (NS, CH, NP) on a single multi-core node. Figures 7 and 8 show the execution time and speedup results for parallelizing the three matchers for up to eight parallel match threads for the medium-scale and large-scale match problems, respectively. We observe that execution times can be significantly improved by increasing the degree of parallelism for all matchers and both match problems. The NP matcher with its long concatenated name strings is by far the most expensive matcher with about four times longer execution times than CH; for the large-scale problem it takes more than 6 hours without parallelism. For the medium-scale match problem NS and CH take about the same time while NS takes much more time for the large-scale problem. This is because GO has many synonyms per concept so that for every concept pair about 11 (instead of 3 in the medium-scale problem) comparisons have to be computed.

For all matchers we achieve excellent speedup values of up to 3.6-4.2 for the medium-scale problem and even 4.5-5 for the large-scale problem. For up to four threads (= number of cores) we achieve almost linear speedup (up to 3.5). Increasing the number of threads brings further improvements (especially for the large-scale problem) but at a reduced level. This is likely because the additional threads can utilize the cores when other match threads are waiting for new tasks to process.

5.3 Parallel Ontology Matching on Multiple Nodes

We now evaluate parallelization strategies using up to four compute nodes (16 cores) running up to four threads per node. In this experiment we combine the three individual matchers NP, CH and NS according to the following parallelization strategies: no

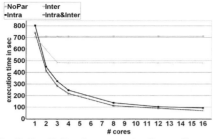
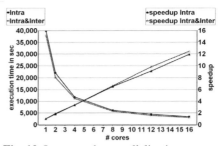

Fig. 9. Parallelization strategies for medium-scale problem

Fig. 10. Intra-matcher parallelization strategies for large scale problem

parallelization *(NoPar)*, inter-matcher parallelization *(Inter)*, intra-matcher parallelization *(Intra)* as well as the combination of both intra- and inter-matcher parallelization *(Intra&Inter)*.

Figure 9 shows the execution time results for these strategies on the medium-scale match problem. *NoPar* is the base case that does not benefit from multiple threads and cores. The other parallelization strategies lead to a performance improvement using more than one core. However, there are differences. *Inter* benefits only to a small degree since we do not apply intra-matcher but only inter-matcher parallelism. Since we apply three matchers we can only improve execution times for up to three cores/threads, i.e., multiple cores are not utilized for our match strategy. The total execution time is limited by the slowest matcher (NP). In contrast, *Intra* and *Intra&Inter* are very effective and achieve matching times of under 100 s. The combined *Intra&Inter* parallelization is slightly better than only using *Intra* and achieves a speedup of up to 10.6 (vs. 8.6). This is because *Intra* executes the three matchers sequentially resulting in some execution delays between matchers that are avoided for the combined approach.

Figure 10 shows the execution time and speedup results for the two parallelization strategies *Intra* and *Intra&Inter* for the large-scale match problem. Due to the large ontology sizes we omit the cases without intra-matcher parallelism and partitioning *(NoPar, Inter)*. The sequential match time for the three matchers is 11h. Using 16 cores *Intra* and *Intra&Inter* reduce the overall execution time to 55 and 50 min and achieve thus an impressive speedup of 11.9 and 12.5, respectively. So, the speedup could be increased compared to the medium-scale match case, similar to the parallelization on a single node. This shows *Intra* and *Intra&Inter* are especially valuable for parallel matching of large ontologies.

6 Related Work

Matching life science ontologies has attracted considerable interest, particularly the matching of anatomy ontologies [22, 33] and molecular biological ontologies [4, 17, 26]. Typically, these studies aim at improving the quality of match results while efficiency aspects found only little attention. The performance of matching large schemas and ontologies in general is considered an open issue [3, 29]. In the past different algorithmic optimizations and fragmentation techniques for improving ontology as well as schema matching performance have been proposed.

Some approaches aim at reducing the search space compared to the Cartesian product for improved performance. Several divide-and-conquer approaches have been proposed where only parts of the input ontologies are matched against each other. [25,7] propose a fragment-based schema matching approach for COMA++ [1] where only similar fragments / sub-schemas need to be matched with each other. [14] partition entities of the input ontologies into sets of clusters and construct blocks which are matched based on pre-calculated anchors. The authors assess that the anchor pre-calculation consumes a main part of the overall runtime. The Anchor-Flood algorithm proposed in [28] also uses anchors (pairs of look-alike concepts) to gradually explore neighboring concepts in order to match only between ontology segments. In [27] nodes are clustered based on a linguistic label similarity and performance can be improved through minimization of the search space.

[23] propose a rule-based optimization technique to rewrite match strategies for improved performance. In particular, newly added filter operators allow a reduction of a matcher output and can thus speedup subsequently executed matchers. QOM [8] uses heuristics to reduce the number of candidate mappings to avoid the complete pair-wise comparison. These candidate mappings are classified into promising and less promising pairs by exploiting the ontological structures.

All these optimizations rely on algorithmic optimizations or partitioning/ fragmentation strategies to reduce the number of comparisons for improved performance. However, these approaches often lead to reduced match quality because relevant correspondences can be missed. Furthermore, the applicability of the approaches is dependent on the considered ontologies and match techniques. In contrast our parallelization strategies are orthogonal and general techniques to improve the performance of matchers and match strategies. They are especially valuable for large-scale match problems. We have shown their usefulness for evaluating the Cartesian product but they should also be usable in combination with other performance optimizations such as reduced search spaces.

7 Conclusion and Future Work

We propose general strategies for parallel ontology matching on multiple compute nodes, namely *inter-* and *intra-matcher parallelization* and their combination. They allow us to execute whole matchers in parallel and to parallelize matchers internally using data partitioning. For intra-matcher parallelism we propose a *size-based partitioning* enabling good load balancing, scalability and limited memory consumption without reducing the quality of match results. We described how element-level, structure-level, instance-based matchers can be parallelized and use multi-valued *context attributes* for structural matching. We implemented a distributed infrastructure that enables parallel ontology matching and evaluated our approach for large life science ontology match problems. The results show the efficiency and scalability for single matchers as well as combined match strategies, especially for large match problems and for the combination of inter- and intra-matcher parallelism.

There are several opportunities for future work. Parallel ontology matching can be investigated for additional matchers. Furthermore, parallelization can be combined with algorithmic performance optimizations and advanced fragmentation strategies proposed in previous work. Moreover, parallel ontology matching may be extended to larger configurations such as cloud infrastructures.

Acknowledgments. This work is supported by the German Research Foundation (DFG), grant RA 497/18-1 ("Evolution of Ontologies and Mappings").

References

1. Aumueller, D., Do, H.H., Massmann, S., Rahm, E.: Schema and ontology matching with COMA++. In: Proc. of ACM SIGMOD Intl. Conference on Management of Data, pp. 906–908 (2005)
2. Bastian, F., Parmentier, G., Roux, J., et al.: Bgee: Integrating and Comparing Heterogeneous Transcriptome Data Among Species. In: Bairoch, A., Cohen-Boulakia, S., Froidevaux, C. (eds.) DILS 2008. LNCS (LNBI), vol. 5109, pp. 124–131. Springer, Heidelberg (2008)
3. Bernstein, P.A., Melnik, S., Petropoulos, M., Quix, C.: Industrial Strength Schema Matching. ACM SIGMOD Record 33(4), 38–43 (2004)
4. Bodenreider, O., Burgun, A.: Linking the Gene Ontology to other biological ontologies. In: Proc. of 8th ISMB Meeting on Bio-Ontologies, pp. 17–18 (2005)
5. Bodenreider, O., Stevens, R.: Bio-ontologies: current trends and future directions. Briefings in Bioinformatics 7(3), 256–274 (2006)
6. Do, H.H., Rahm, E.: COMA – A System for Flexible Combination of Schema Matching Approaches. In: Proc. of the 28th Intl. Conference on Very Large Databases (VLDB), pp. 610–621 (2002)
7. Do, H.H., Rahm, E.: Matching large schemas: Approaches and evaluation. Information Systems 32(6), 857–885 (2007)
8. Ehrig, M., Staab, S.: QOM – Quick Ontology Mapping. In: McIlraith, S.A., Plexousakis, D., van Harmelen, F. (eds.) ISWC 2004. LNCS, vol. 3298, pp. 683–697. Springer, Heidelberg (2004)
9. Euzenat, J., Shvaiko, P.: Ontology Matching. Springer, Heidelberg (2007)
10. The Gene Ontology Consortium: The Gene Ontology project in 2008. Nucleic Acids Research 36(Database issue), D440–D444 (2008)
11. Gross, A., Hartung, M., Kirsten, T., Rahm, E.: Estimating the Quality of Ontology-Based Annotations by Considering Evolutionary Changes. In: Paton, N.W., Missier, P., Hedeler, C. (eds.) DILS 2009. LNCS (LNBI), vol. 5647, pp. 71–87. Springer, Heidelberg (2009)
12. Hartung, M., Kirsten, T., Rahm, E.: Analyzing the Evolution of Life Science Ontologies and Mappings. In: Bairoch, A., Cohen-Boulakia, S., Froidevaux, C. (eds.) DILS 2008. LNCS (LNBI), vol. 5109, pp. 11–27. Springer, Heidelberg (2008)
13. Hayamizu, T.F., Mangan, M., Corradi, J.P., Kadin, J.A., Ringwald, M.: The Adult Mouse Anatomical Dictionary: a tool for annotating and integrating data. Genome Biology 6(3), R29 (2005)
14. Hu, W., Qu, Y., Cheng, G.: Matching large ontologies: A divide-and-conquer approach. Data & Knowledge Engineering 67(1), 140–160 (2008)
15. Jakoniene, V., Lambrix, P.: Ontology-based integration for bioinformatics. In: Proc. VLDB Workshop on Ontologies-based techniques for Databases and Information Systems (ODBIS), pp. 55–58 (2005)

16. Kirsten, T., Hartung, M., Gross, A., Rahm, E.: Efficient Management of Biomedical Ontology Versions. In: Meersman, R., Herrero, P., Dillon, T.S. (eds.): On the Move to Meaningful Internet Systems Workshops. Proceedings. LNCS, vol. 4544, pp. 172-187. Springer, Heidelberg (2007)

17. Kirsten, T., Thor, A., Rahm, E.: Instance-based matching of large life science ontologies. In: Bairoch, A., Cohen-Boulakia, S., Froidevaux, C. (eds.) DILS 2008. LNCS (LNBI), vol. 5109, pp. 11–27. Springer, Heidelberg (2008)

18. Lambrix, P., Edberg, A.: Evaluation of ontology merging tools in bioinformatics. In: Proc. of the 8th Pacific Symposium on Biocomputing, pp. 589–600 (2003)

19. Lambrix, P., Tan, H., Jakoniene, V., Strömbäck, L.: Biological Ontologies. In: Semantic Web: Revolutionizing Knowledge Discovery in the Life Sciences, pp. 85–99 (2007)

20. Ontology Alignment Evaluation Initiative, http://20.ontologymatching.org/

21. Melnik, S., Garcia-Molina, H., Rahm, E.: Similarity Flooding: A Versatile Graph Matching Algorithm and Its Application to Schema Matching. In: Proc. of the 18th Intl. Conference on Data Engineering (ICDE), pp. 117–128 (2002)

22. Mork, P., Bernstein, P.A.: Adapting a Generic Match Algorithm to Align Ontologies of Human Anatomy. In: Proc. of the 20th Intl. Conference on Data Engineering (ICDE), pp. 787–790 (2004)

23. Peukert, E., Berthold, H., Rahm, E.: Rewrite Techniques for performance Optimization of Schema Matching Processes. In: Proc. 13th Intl. Conference on Extending Database Technology (EDBT), pp. 453–464 (2010)

24. Rahm, E., Bernstein, P.A.: A survey of approaches to automatic schema matching. VLDB Journal 10(4), 334–350 (2001)

25. Rahm, E., Do, H.H., Massmann, S.: Matching large XML schemas. ACM SIGMOD Record 33(4), 26–31 (2004)

26. Rance, B., Gibrat, J.F., Froidevaux, C.: An Adaptive Combination of Matchers: Application to the Mapping of Biological Ontologies for Genome Annotation. In: Paton, N.W., Missier, P., Hedeler, C. (eds.) DILS 2009. LNCS (LNBI), vol. 5647, pp. 113–126. Springer, Heidelberg (2009)

27. Saleem, K., Bellahsene, Z., Hunt, E.: PORSCHE: Performance ORiented SCHEma mediation. Information Systems 33(7-8), 637–657 (2008)

28. Seddiqui, H., Aono, M.: An efficient and scalable algorithm for segmented alignment of ontologies of arbitrary size. Web Semantics: Science, Services and Agents on the World Wide Web 7(4), 344–356 (2009)

29. Shvaiko, P., Euzenat, J.: Ten challenges for ontology matching. In: Proc. of on the Move to Meaningful Internet Systems (OTM), pp. 1164–1182 (2008)

30. Sioutos, N., de Coronado, S., Haber, M.W., et al.: NCI Thesaurus: A semantic model integrating cancer-related clinical and molecular information. Journal of Biomedical Informatics 40(1), 30–43 (2007)

31. Smith, B., Ashburner, M., Rosse, C., et al.: The OBO Foundry: coordinated evolution of ontologies to support biomedical data integration. Nature Biotechnology 25(11), 1251–1255 (2007)

32. Thor, A., Hartung, M., Gross, A., Kirsten, T., Rahm, E.: An evolution-based approach for assessing ontology mappings - A case study in the life sciences. In: Proc. Conference of the Business, Technology and Web (BTW), pp. 277–286 (2009)

33. Zhang, S., Bodenreider, O.: Aligning Representations of Anatomy using Lexical and Structural Methods. In: Proc. of AMIA Annual Symposium, pp. 753–757 (2003)

A System for Debugging
Missing Is-a Structure in Networked Ontologies

Qiang Liu and Patrick Lambrix

Department of Computer and Information Science
Linköpings universitet, 581 83 Linköping, Sweden

Abstract. Ontologies are recognized as a key technology for semantics-based integration of the many available biomedical data sources. However, developing ontologies is not an easy task and the resulting ontologies may have defects affecting the results of ontology-based data integration and retrieval. In this paper we present a system for debugging ontologies regarding an important kind of modeling defects. Our system supports a domain expert to detect and repair missing is-a structure in ontologies in a semi-automatic way. The input for our system is a set of ontologies networked by correct mappings between their terms. Our tool uses the ontologies and mappings as domain knowledge to detect missing is-a relations in these ontologies. It also assists the user in repairing the ontologies by generating and recommending possible ways of repairing and executing the chosen repairing strategy. The detection and repairing phases can be interleaved. We present our approach, an implemented system as well as an experiment with two anatomy ontologies.

1 Introduction

The success of the large data generation projects in the Life Sciences in combination with the popularity of the World Wide Web have made a large amount of biomedical data available to the scientific community through the Internet [2]. The resulting data sources are heterogeneous in different ways and how to integrate this heterogenous data has become one of the most challenging problems facing bioinformatics today [11]. To deal with this, ontologies are recognized as a key technology. Intuitively, ontologies can be seen as defining the basic terms and relations of a domain of interest, as well as the rules for combining these terms and relations [21]. As a basis for interoperability between data sources, ontologies facilitate information reuse, sharing and portability across platforms, and improved documentation, maintenance, and reliability. The work on ontologies is recognized as essential in some of the grand challenges of genomics research [4] and there is much international research cooperation for the development of ontologies (many of which are available from Open Biological and Biomedical Ontologies (OBO) [23]) and the use of ontologies for data source annotation, and data search, integration and exchange [15].

Developing ontologies is not an easy task. When the ontologies grow in size containing thousands to tens of thousands of terms, it is difficult to ensure the correctness and completeness of the ontologies. The resulting ontologies may be not consistent or structurally complete. For instance, in [13] it was shown that for the two real-world ontologies

P. Lambrix and G. Kemp (Eds.): DILS 2010, LNBI 6254, pp. 50–57, 2010.

used in the Anatomy track in the 2008 and 2009 Ontology Alignment Evaluation Initiative (OAEI), Adult Mouse Anatomy Dictionary [16] (MA, 2744 concepts) and the NCI Thesaurus anatomy [20] (NCI-A, 3304 concepts), at least 121 is-a relations in MA and 83 in NCI-A are missing. This is not an uncommon case. It is well-known that people that are not expert in knowledge representation often misuse and confuse equivalence, is-a and part-of (e.g. [5]), which leads to problems in the structure of the ontologies.

Such ontologies, although often useful, also cause problems for the intended use. Wrong conclusions may be derived or valid conclusions may be missed. For instance, the defect of incomplete structure in ontologies influences ontology-based search, in which queries are refined and expanded by moving up and down the hierarchy of concepts. As an example, suppose we want to find articles in MeSH (Medical Subject Headings [19]) Database of PubMed [24] using the term *Scleral Diseases* in MeSH. By default the query will follow the hierarchy of MeSH and include more specific terms for searching, such as *Scleritis*. If the relation between *Scleral Diseases* and *Scleritis* is missing in MeSH, we will miss 738 articles (about 55% of the original result) in the search result.

To deal with ontological defects, we need to debug ontologies [26], i.e. detect missing and wrong information and then repair the ontologies. Up to date most work has been performed on debugging the semantic defects such as unsatisfiable concepts and inconsistent ontologies (e.g. [25,10,18,9]). Detecting and resolving modeling defects requires, in contrast to semantic defects, the use of domain knowledge. One interesting kind of domain knowledge are the other ontologies and information about connections between these ontologies. For instance, in the case of the Anatomy track in OAEI, we were able to detect the missing is-relations by using MA and NCI-A as domain knowledge for each other together with a partial reference alignment (PRA, a set of correct mappings between the terms of the ontologies) containing 988 mappings. Recently, more and more mappings are being produced and thus more and more ontologies are being networked, and a number of systems and portals have been set up that store mappings between ontologies (e.g. Unified Medical Language System (UMLS) [27], BioPortal [22]).

Once the missing is-a relations are found, the ontology can be repaired by adding a set of is-a relations (called a *structural repair* in [14]) such that when these are added,

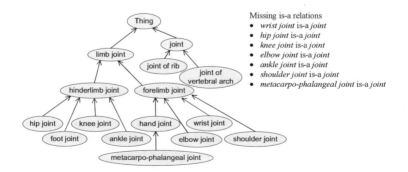

Fig. 1. A part of MA regarding the concept *joint*

all missing is-a relations can be derived from the extended ontology. Clearly, the missing is-a relations themselves constitute a structural repair, but this is not always the most interesting solution for a domain expert. For instance, Figure 1 shows a part of MA regarding the concept *joint* (is-a relations shown with arrows). Using NCI-A and the PRA as domain knowledge, 7 missing is-a relations are found. These missing is-a relations themselves could be a structural repair. However, for the missing is-a relation "*wrist joint* is-a *joint*", knowing that there is an is-a relation between *wrist joint* and *limb joint*, a domain expert will most likely prefer to add the is-a relation "*limb joint* is-a *joint*" instead. This is correct from a modeling perspective as well as more informative and would lead to the fact that the missing is-a relation between *wrist joint* and *joint* can be derived. In this particular case, using "*limb joint* is-a *joint*" would actually also lead to the repairing of the other 6 missing is-a relations, as well as others that were not found before (e.g. "*hand joint* is-a *joint*"). In general, such a decision should be made by domain experts.

2   Related Work

There is not much work on detecting and repairing modeling defects in networked ontologies. In [1] and [13] similar strategies to detect missing is-a relations are described. Given two pairs of terms between two ontologies which are linked by the same kind of relationship, if the two terms in one ontology are linked by an is-a relation while the corresponding terms in the other are not, it is deemed as a possible missing is-a relation. The preliminary work of this paper is in [14], where we presented a system that supports the repairing of the missing is-a structure in a single ontology when some missing is-a relations are known.

Related to the detection of missing relations, there is much work on finding relationships between terms in a single ontology in the text mining area. Much of the work on detecting is-a relations is based on the use of Hearst patterns [7] or extensions thereof (e.g. [3,28]). Most of these approaches have good precision, but low recall. A semiautomatic approach for ontology refinement (including is-a relations) is given in [29]. In [30] it was shown that superstring prediction and co-occurrence analysis may be used to detect is-a relations. In [32] a statistical approach is used. An overview of approaches in ontology learning is given in [17].

3   Overview of the Debugging Approach

In this section, we give an overview of our debugging approach. For the theory behind our approach as well as details about the algorithms we refer to [12].[1] As illustrated in Figure 2, the whole process consists of 5 phases and is driven by a domain expert.

The input is a set of ontologies networked by a set of PRAs which contain equivalence and subsumption axioms. First, missing is-a relations for all the ontologies in the network are detected (**Phase 1**). The intuition is that, for each ontology, if there is an is-a relation between a pair of concepts not derivable from the ontology alone, but

---

[1] This work is an extension of the work presented in [14].

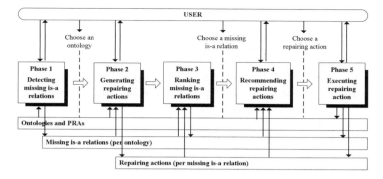

Fig. 2. Approach for debugging missing is-a structure in networked ontologies

derivable from the ontological network, it is identified as a missing is-a relation. This is how we found the 7 missing is-a relations in Figure 1.

After this, the user can choose an ontology and generate possible ways of repairing, called *repairing actions*, for all missing is-a relations in the ontology (**Phase 2**). The resulting repairing actions are presented as two sets of concepts, called *Source* and *Target* sets. A possible repairing action is an is-a relation "A is-a B" where A is an element from the Source set and B is an element from the Target set. Any pair from Source x Target would allow us, when added to the ontology, to derive the missing is-a relation. As an example, for the missing is-a relation *"wrist joint* is-a *joint"* in Figure 1, the Source and Target sets will be {*wrist joint, forelimb joint, limb joint*} and {*joint, joint of vertebral arch, joint of rib*} respectively, which give 9 possible repairing actions. For the computation of possible repairing actions, we have implemented two algorithms that implement three heuristics. The first heuristic prefers not to use non-contributing is-a relations for repairing. The second heuristic prefers to use the most informative repairing actions. The third heuristic prefers not to change is-a relations in the original ontology into equivalence relations. For details we refer to [14]. In practice, there will be many missing is-a relations that need to be repaired and some of them may be easier to start with such as the ones with fewer repairing actions. We therefore rank them with respect to the number of possible repairing actions (**Phase 3**).

Then the user can select a missing is-a relation to repair and choose between possible repairing actions. To facilitate this process, we developed a method to recommend the most informative repairing actions based on domain knowledge (**Phase 4**). In the previous example, the recommended repairing action for the missing is-a relation *"wrist joint* is-a *joint"* given by WordNet [31] is *"limb joint* is-a *joint"*.

Once the user chooses a repairing action to execute, the chosen repairing action is then added to the ontology and the consequences are computed (**Phase 5**). Some other missing is-a relations may be repaired by the executed repairing action, such as in the case in Figure 1 when the repairing action *"limb joint* is-a *joint"* is executed. For some other missing is-a relations, the Source and Target sets may change. Further, some new missing is-a relations in ontologies may be found.

At any time during the process, the user can switch the ontology to repair or start earlier phases.

4   Implemented System

We implemented our system RepOSE (*Rep*air of *O*ntological *S*tructure *E*nvironment) based on the approach described in Section 3. We use a framework and reasoner provided by Jena (version 2.5.7) [8]. The domain knowledge that we use includes WordNet and UMLS. Here, we show its use using pieces of MA and NCI-A regarding the concept *joint*, as well as a PRA with 8 equivalence mappings.

As input our system takes a set of ontologies in OWL format as well as a set of PRAs in RDF format. The ontologies and PRAs can be imported using the *Load Ontologies and PRAs* button. The user can see the list of ontologies in the *Ontologies* menu (see Figure 3). Once the *Detect Missing IS-A Relations* button is clicked, missing is-a relations are detected in all ontologies. Then, the user can select which ontology to repair, and the *Missing IS-A Relations* menu shows the missing is-a relations of the currently selected ontology. In this case the ontology *joint_mouse_anatomy.owl* is selected and it contains 7 missing is-a relations (same as the case in Figure 1).

Clicking on the *Generate Repairing Actions* button, results in the computation of repairing actions for the missing is-a relations of the ontology under repair, which is preceded by a two-stage preprocessing step. During the preprocessing, one stage is to identify the missing is-a relations which are actually equivalence relations and repair them by adding the equivalence relations. The other is to identify and remove the redundant missing is-a relations which are derivable from the ontology extended with other missing is-a relations. Then, repairing actions for each missing is-a relation are computed and presented as Source and Target sets. The selection of the *useExtendAlg* checkbox makes the computation use our extended algorithm, otherwise our basic algorithm is used. Once the Source and Target sets are computed, the missing is-a relations

Fig. 3. User interface of RepOSE

are ranked with respect to the number of possible repairing actions. The first missing is-a relation in the list has the fewest possible repairing actions, and may therefore be a good starting point. When the user chooses a missing is-a relation, the Source and Target sets for the repairing actions are shown in the panels on the left and the right, respectively (as shown in Figure 3). Both these panels have zoom control and could be opened in a separate window by double clicking. The concepts in the missing is-a relation are highlighted in red. In this case, the repairing actions of the missing is-a relations are generated using the basic algorithm. The selection of the missing is-a relation "*wrist joint* is-a *joint*" displays its Source and Target sets in the panels. They contain 3 and 26 concepts respectively.

For the selected missing is-a relation, the user can also ask for recommended repairing actions by clicking the *Recommend Repairing Actions* button. The two checkboxes allow the user to specify the external domain knowledge used for generating recommendations. In our case, the system uses *WordNet* and recommends to add an is-a relation between *limb joint* and *joint*. In general, the system presents a list of recommendations. By selecting an element in the list, the concepts in the recommended repairing action are identified by round boxes in the panels. The user can repair the missing is-a relation by selecting a concept in the Source panel and a concept in the Target panel and clicking on the *Repair* button. The repairing action is then added to the ontology, and other missing is-a relations are updated, as well as the set of missing is-a relations of every ontology in the network.

At all times during the process the user can inspect the ontology under repair by clicking the *Show Ontology* button. The is-a structure of the repaired ontology will be shown in a separate window with newly added is-a relations being highlighted. The user can save the repaired ontology into an OWL file by clicking the *Save* button, or select another ontology to repair. The whole debugging process runs semi-automatically until no more missing is-a relations are found or unrepaired in the networked ontologies.

5   Discussion

We tested the feasibility of our approach using MA and NCI-A. After loading the two ontologies and the PRA, our system found 199 missing is-a relations in MA and 167 in NCI-A during the initial detection phase. These missing is-a relations are preprocessed before the computation of the repairing actions. For MA, 6 missing equivalence relations are identified and repaired immediately, while 74 redundant missing is-a relations are found and removed. For NCI-A, the numbers of missing equivalence relations and redundant missing is-a relations are 3 and 84 respectively. As a result, we have 119 missing is-a relations to repair in MA and 80 in NCI-A after the preprocessing. As for the computation of repairing actions, for MA, our basic algorithm generates for 9 missing is-a relations only 1 repairing action (which is then the missing is-a relation itself). Therefore these could be immediately repaired. For NCI-A this number is 5. Of the remaining missing is-a relations there are 64 missing is-a relations for MA that have only 1 element in the Source set and 2 missing is-relations that have 1 element in the Target set. For NCI-A these numbers are 20 and 3, respectively. These are likely to be good starting points for repairing. For most of the missing is-a relations the Source and Target sets are small and thus can be easily visualized in the panels of our system.

After this, we run the repairing session completely. As we are not domain experts, we used [6] to decide on possible choices along with the recommendation algorithm based on WordNet. Clearly, we aim to redo this experiment with domain experts. However, this run has given us some interesting information. It took about 3 hours to repair these two ontologies. During the process, we found 6 new missing is-a relations in MA and 10 in NCI-A by repairing other is-a relations. In most cases the recommendations seemed useful. For NCI-A the system recommended repairing actions other than the missing is-a relation itself, for only 5 missing is-a relations and each of these received 1 recommended repairing action. For MA 23 missing is-a relations received 1 recommended repairing action, 11 received 2 and 2 received 3. For 27 missing is-a relations in MA and 10 in NCI-A the Target set was too large to have a good visualization in the tool.

6   Conclusion

In this paper we presented a system for debugging the missing is-a structure in networked ontologies. We proposed an approach, developed algorithms and implemented a system that allows a domain expert to detect and repair the is-a structure of ontologies in a semi-automatic way.

There are a number of directions that are interesting for future work. Since this work uses PRAs as domain knowledge assuming that the given mappings are correct, a direct extension is the case when these mappings are not necessarily correct. In this case, we will need to also deal with the repairing of the mappings (semantic defects) such as in [9] and [18]. Another interesting direction is to deal with ontologies represented in more expressive representation languages, and investigate possible influences between semantic defects and modeling effects.

References

1. Bada, M., Hunter, L.: Identification of OBO nonalignments and its implication for OBO enrichment. Bioinformatics 24(12), 1448–1455 (2008)
2. Cheung, K.-H., Smith, A., Yip, K., Baker, C., Gerstein, M.: Semantic web approach to database integration in the life sciences. In: Baker, Cheung (eds.) Semantic Web: revolutionizing knowledge discovery in the life sciences, pp. 11–30. Springer, Heidelberg (2007)
3. Cimiano, P., Staab, S.: Learning by googling. ACM SIGKDD Explorations Newsletter 6(2), 24–33 (2004)
4. Collins, F., Green, E., Guttmacher, A., Guyer, M.: A vision for the future of genomics research. Nature 422, 835–847 (2003)
5. Conroy, C., Brennan, R., O'Sullivan, D., Lewis, D.: User evaluation study of a tagging approach to semantic mapping. In: Aroyo, L., Traverso, P., Ciravegna, F., Cimiano, P., Heath, T., Hyvönen, E., Mizoguchi, R., Oren, E., Sabou, M., Simperl, E. (eds.) ESWC 2009. LNCS, vol. 5554, pp. 623–637. Springer, Heidelberg (2009)
6. Feneis, F., Dauber, W.: Pocket Atlas of Human Anatomy, 4th edn. Thieme Verlag (2000)
7. Hearst, M.A.: Automatic acquisition of hyponyms from large text corpora. In: 14th Int. Conf. on Computational Linguistics, pp. 539–545 (1992)
8. Jena, http://jena.sourceforge.net/
9. Ji, Q., Haase, P., Qi, G., Hitzler, P., Stadtmuller, S.: RaDON - repair and diagnosis in ontology networks. In: Aroyo, L., et al. (eds.) ESWC 2009. LNCS, vol. 5554, pp. 863–867. Springer, Heidelberg (2009)

10. Kalyanpur, A., Parsia, B., Sirin, E., Hendler, J.: Debugging unsatisfiable classes in OWL ontologies. Journal of Web Semantics 3(4), 268–293 (2006)
11. Lacroix, Z., Critchlow, T.: Bioinformatics: managing scientific data. Morgan Kaufmann, San Francisco (2003)
12. Lambrix, P., Liu, Q.: Debugging the missing is-a structure within ontologies networked by partial reference alignments (forthcoming)
13. Lambrix, P., Liu, Q.: Using partial reference alignments to align ontologies. In: Aroyo, L., Traverso, P., Ciravegna, F., Cimiano, P., Heath, T., Hyvönen, E., Mizoguchi, R., Oren, E., Sabou, M., Simperl, E. (eds.) ESWC 2009. LNCS, vol. 5554, pp. 188–202. Springer, Heidelberg (2009)
14. Lambrix, P., Liu, Q., Tan, H.: Repairing the missing is-a structure of ontologies. In: 4th Asian Semantic Web Conf., pp. 371–386 (2009)
15. Lambrix, P., Strömbäck, L., Tan, H.: Information integration in bioinformatics with ontologies and standards. In: Bry, Maluszynski (eds.) Semantic Techniques for the Web: The REWERSE perspective, pp. 343–376. Springer, Heidelberg (2009)
16. MA. Adult mouse anatomical dictionary,
    http://www.informatics.jax.org/searches/AMA_form.shtml
17. Maedche, A., Pekar, V., Staab, S.: Ontology learning part one - on discovering taxonomic relations from the web. In: Zhong, Liu, Yao (eds.) Web Intelligence, pp. 301–322. Springer, Heidelberg (2003)
18. Meilicke, C., Stuckenschmidt, H., Tamilin, A.: Repairing ontology mappings. In: 20th National Conf. on Artificial Intelligence, pp. 1408–1413 (2007)
19. MeSH. Medical subject headings, http://www.nlm.nih.gov/mesh/
20. NCI-A. National cancer institute - anatomy,
    http://www.cancer.gov/cancerinfo/terminologyresources/
21. Neches, R., Fikes, R., Finin, T., Gruber, T., Patil, R., Senator, T., Swartout, W.: Enabling technology for knowledge sharing. AI Magazine 12(3), 36–56 (1991)
22. Noy, N.F., Griffith, N., Musen, M.: Collecting community-based mappings in an ontology repository. In: 7th Int. Semantic Web Conf., pp. 371–386 (2008)
23. OBO. Open biological and biomedical ontologies, http://obo.sourceforge.net/
24. PubMed, http://www.ncbi.nlm.nih.gov/pubmed/
25. Schlobach, S.: Debugging and semantic clarification by pinpointing. In: Gómez-Pérez, A., Euzenat, J. (eds.) ESWC 2005. LNCS, vol. 3532, pp. 226–240. Springer, Heidelberg (2005)
26. Schlobach, S., Cornet, R.: Non-standard reasoning services for the debugging of description logic terminologies. In: 18th Int. Joint Conf. on Artificial Intelligence, pp. 355–360 (2003)
27. UMLS. Unified medical language system,
    http://www.nlm.nih.gov/research/umls/about_umls.html
28. van Hage, W.R., Katrenko, S., Schreiber, G.: A method to combine linguistic ontology-mapping techniques. In: Gil, Y., Motta, E., Benjamins, V.R., Musen, M.A. (eds.) ISWC 2005. LNCS, vol. 3729, pp. 732–744. Springer, Heidelberg (2005)
29. Völker, J., Hitzler, P., Cimiano, P.: Acquisition of OWL DL axioms from lexical resources. In: Franconi, E., Kifer, M., May, W. (eds.) ESWC 2007. LNCS, vol. 4519, pp. 670–685. Springer, Heidelberg (2007)
30. Wächter, T., Tan, H., Wobst, A., Lambrix, P., Schroeder, M.: A corpus-driven approach for design, evolution and alignment of ontologies. In: Winter Simulation Conf., pp. 1595–1602 (2006) (invited contribution)
31. WordNet, http://wordnet.princeton.edu/
32. Zavitsanos, E., Paliouras, G., Vouros, G.A., Petridis, S.: Discovering subsumption hierarchies of ontology concepts from text corpora. In: IEEE/WIC/ACM Int. Conf. on Web Intelligence, pp. 402–408 (2007)

On the Secure Sharing and Aggregation of Data to Support Systems Biology Research

Andrew Simpson[1], Mark Slaymaker[1,2], and David Gavaghan[1,2]

[1] Oxford University Computing Laboratory
Wolfson Building
Parks Road
Oxford OX1 3QD
United Kingdom
[2] Oxford Centre for Integrative Systems Biology
Department of Biochemistry
South Parks Road
Oxford OX1 3QU
United Kingdom

Abstract. The development of tools and technologies to facilitate appropriate and effective data sharing is becoming increasingly important in many academic disciplines. In particular, the 'data explosion' problem associated with the Life Sciences has been recognised by many researchers and commented upon widely, as have the associated data management problems. In this paper we describe how a middleware framework that supports the secure sharing and aggregation of data from heterogeneous data sources—developed initially to underpin the sharing of healthcare-related data—is being used to support Systems Biology research at the University of Oxford. As well as giving an overview of the framework and its application, we attempt to set our work within the wider context of the emerging challenges associated with data sharing within the Life Sciences.

1 Introduction

The emerging data challenges facing researchers in the Life Sciences—in terms of, for example, capture, storage and curation—are significant, and it is well understood that the pressing need to develop appropriate tools and technologies to facilitate effective data management is likely to become increasingly urgent in the coming years.

These challenges have, of course, been recognised and commented upon by many authors; for example, to quote Kim [1]:

> "Modern large-scale data collection efforts require fundamental infrastructure support for archiving data, organizing data into structured information (e.g., data models and ontologies), and disseminating data to the broader community. Furthermore, distributed data collection efforts require coordination and integration of the heterogeneous data resources."

P. Lambrix and G. Kemp (Eds.): DILS 2010, LNBI 6254, pp. 58–73, 2010.

A further challenge—and one with which we are fundamentally concerned in this paper—involves determining how such data might be *shared* effectively and appropriately.

The Oxford Centre for Integrative Systems Biology (OCISB)[1] was established to "strengthen existing and forge more interdisciplinary collaborations in the pursuit of joint experimental and theoretical research in Systems Biology"—with data sharing being at the heart of such collaborations. The Centre brings together the expertise of 17 investigators from 9 different departments, and currently employs 17 post-doctoral researchers. In this paper we report upon how a middleware framework that facilitates the integration of heterogeneous data resource (*à la* Kim) and supports appropriate data sharing via a fine-grained authorisation mechanism is being utilised within OCISB to support collaboration between researchers from different disciplines.

The middleware framework, *sif* (for service-oriented interoperability framework) [2], which is based on Java and web services, was developed initially within the context of the GIMI (Generic Infrastructure for Medical Informatics) project [3,4] as a means of facilitating the secure sharing of medical data (see, for example, [5] and [6]). However, sif's generic nature—it provides secure access to, and aggregation of, data from any structured source—means that it has been used in a variety of other contexts, including, for example, the integration of student administration data within the University of Oxford.

Importantly, we wish to hide issues of heterogeneity from the end-user. To this end, we utilise the classifications of Ouksel and Sheth [7], where:

- *system* heterogeneity is concerned with the combination of software and hardware associated with a data source;
- *syntactic* heterogeneity is concerned with the low-level encoding of data;
- *structural* heterogeneity is concerned with representation of data; and
- *semantic* heterogeneity is concerned with the meaning and interpretation of data.

sif protects application developers and end-users from issues of systems and syntactic heterogeneity; issues of structural and semantic heterogeneity are also hidden from end-users—but the responsibility for their resolution resides with application developers, who are, typically, domain experts.

Our particular concern in this paper is not sif *per se*—the framework is described in detail elsewhere[2]—but its application to supporting collaboration within the Systems Biology context. To this end, we report upon its support for a particular application within OCISB, in which codes and simulation data are shared between cancer modellers.

The structure of the remainder of the paper is as follows. In Section 2 we describe the motivation for our work, giving particular focus to issues of data

---

[1] See http://www.sysbio.ox.ac.uk/

[2] For example, sif's support for federation is described in [8]; its support for fine-grained access control is described in [9]; and sif's 'plug-in' mechanism—which gives rise to sif's data agnosticism, therefore resolving issues of system and syntactic heterogeneiry—is described in [10].

sharing and integration within the Life Sciences. In Section 3 we provide a necessarily brief overview of our middleware framework, sif. In Section 4 we describe how sif has been utilised to support the aforementioned application, and also describe briefly a second application. Finally, in Section 5 we summarise the contribution of this paper and give an overview of our immediate areas of future work.

2   Motivation

Our primary concerns are data *integration*—facilitating the aggregation of data from disparate data sources[3]—and data *sharing*—ensuring that those researchers with appropriate credentials and permissions can access relevant research data to enable collaboration. A consideration of some the relevant issues in this respect is given by Ives [13]:

> "One of the open challenges . . . lies in developing the right *architectures and models* for supporting effective data integration and exchange in science."

Ives goes further, in arguing:

> "Clearly, an enterprise-oriented view of data integration is mismatched for the needs of the life sciences. We instead need a data sharing scheme that:
> - Accommodates multiple, community-specific schemas and vocabularies that may evolve over time.
> - Supports highly dynamic, repeatedly revised, frequently annotated data.
> - Facilitates sharing across different communities in a way that *scales with the amount of invested effort*: limited data sharing should be easy, and further time investment should enable greater data sharing.
> - Tolerates disagreement among different communities about data items (hypothesized facts).
> - Restricts the exchange of data based on assessments of *source authority* and *mapping quality.*
> - Allows end users to integrate data *across* individual data sources without understanding SQL or schema mappings—but takes into account the query author's perception of the authority or relevance of specific databases."

---

[3] An authoritative (if a little dated) survey of the challenges of integrating data in the Life Sciences is given in [11], while [12] provides an excellent overview of the issues pertaining to integrating data in contexts in which a global data schema is present. Our approach assumes that a global schema data *is not* present—although this is not disallowed, with the task of integration being simplified considerably if one does exist.

This view of lightweight mechanisms to facilitate integration is supported by Paton in [14], in which it is noted that there is "increasing interest in approaches with reduced up-front costs."

Fundamental to our approach—and this is consistent with the arguments of both [13] and [14]—is the assumption of a 'bottom-up' philosophy[4] with respect to the construction of virtual organisations: there is no assumption of a global data schema. This approach is, in other contexts, termed a *peer-to-peer* approach [16]: peer-to-peer systems such as Piazza [17], PeerDB [18] and Orchestra [15] all allow queries to be formulated on one peer and subsequently propagated.

The relevance, and, indeed, the appropriateness, of such a bottom-up approach is argued in [15]:

> "In bioinformatics today there are many 'standard' schemas rather than a single one, due to different research needs, competing groups, and the continued emergence of new kinds of data. These efforts not only fail to satisfy the goal of integrating all of the data sources needed by biologists, but they result in standards that must repeatedly be revised, inconsistencies between different repositories, and an environment that restricts an independent laboratory or scientist from easily contributing 'nonstandard' data.

> "The central problem is that science evolves in a 'bottom-up' fashion, resulting in a fundamental mismatch with top-down data integration methods."

An excellent distinction between what might be considered 'top-down' and 'bottom-up' approaches is given in [19], in which data integration approaches in the Life Sciences are classified into two broad types.

The first type includes

> "projects that achieve a high standard of quality in the integrated data through manual curation, i.e., using the experience and expertise of trained professionals. We call this type of projects 'data-focused'. A prominent example for this class of integrated systems is Swiss-Prot [20], collecting and integrating data on protein sequences from journal publications, submissions, personal communications, and other databases by means of approximately two dozen human data curators. Data-focused projects are typically managed by domain experts, e.g., biologists. Database technology plays an only minor role. All effort is put into acquiring and curating the actual data that is usually maintained in a text-like manner. If detailed schemata are developed and used, they are not exposed to the user for structured queries."

---

[4] It should be noted that [15] uses the same term.

And the second type includes

> "projects, which we call 'schema-focused', [which] are aimed at pro-
> viding integration middleware rather than building concrete databases.
> They mostly deal with schema information, using techniques such as
> schema integration, schema mapping, and mediator-based query rewrit-
> ing ... [they] require for each integrated data source the creation of
> some sort of wrapper for query processing, and a detailed semantic map-
> ping between the heterogeneous source schemata and a global, mediated
> schema."

Our approach fits into this second category. For example, taking a simplified view
(and leaving aside higher-level concerns such as semantics), one might characterise
data interoperability as facilitating both *database interoperability* (between Dr
Smith's breast cancer research database in San Francisco and Dr Thomas' colorectal cancer research database in New York) and *database management system
interoperability* (between the IBM DB2 database utilised by Dr Smith and the Oracle database utilised by Dr Thomas). Our concern is the latter; issues of semantic interoperability are left to application developers. Application developers then
only need to worry about interoperability between relevant data sources—rather
than worrying about interoperability across the whole virtual organisation.

To reprise an example from [2], suppose, say, that, data source S1 might
contain data and files pertaining to both breast and colorectal cancer, data
source S2 might contain data and files pertaining to breast cancer, and data
source S3 might contain data and files pertaining to colorectal cancer. S1 and S2
might form one virtual organisation that is concerned with breast cancer; S1 and
S3 might form a second virtual organisation that is concerned with colorectal

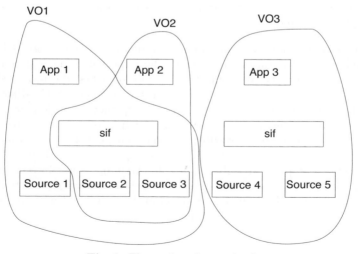

Fig. 1. Three virtual organisations

cancer. Each virtual organisation, then, would be concerned with facilitating semantic interoperability to share relevant data: breast cancer in the case of the first virtual organisation, and colorectal cancer in the case of the second virtual organisation. If, at a later date, the two virtual organisations were to merge to form a single community of interest, then, at that point, issues of interoperability between the breast and colorectal cancer data sets would have to be considered.

Such a state of affairs is illustrated in Figure 1: VO1 involves three data sources, accessed via Application 1; VO2 involves two data sources (two of which also participate in VO1), accessed via Application 2; and VO3 exists in isolation from VO1 and VO2.

3   sif: A Service-Oriented Framework for the Secure Sharing and Aggregation of Data

In this section we provide a necessarily brief overview of our middleware framework, sif (service-oriented interoperability framework).

sif is fundamentally concerned with supporting 'big ideas'—bigger and better research; personalised healthcare; joined up e-government—but in a way that doesn't require organisations to throw away existing systems, change practices, or invest heavily in new technology.

Our drivers can, therefore, be characterised in terms of:

- interoperability, heterogeneity and portability: any kind of data stored on any kind of database or file system should be capable of being accessed and shared via a standard interface;
- secure data sharing: data access and transfer should be in accordance with the data owners' wishes, no matter how prescriptive;
- low costs of entry—in terms of installation and deployment, system footprint and effort required on behalf of application developers; and
- abstraction: via a simple API, application developers can construct applications to aggregate and utilise data without concerning themselves with issues such as secure data transport.

The philosophy behind sif was originally described in [21]. There, a virtual organisation—spread across two or more geographically or physically distinct units—was characterised in terms of the diagram of Figure 2. Deployments communicate via their external interfaces (represented by E), with data being accessed via an internal interface, I. The permitted access to the data is regulated by policies, represented by P. Key to this representation is the concept that each organisation has ultimate control over the access to the data it holds—when sharing data within the context of a virtual organisation, this allows a data provider to share only the data that the data owner wishes. This leaves the responsibility for defining the policies associated with the access to data to the institution that owns the data (although the deployment of 'top-down' global policies can be supported if necessary).

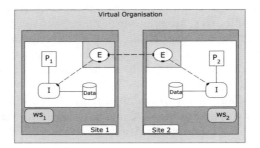

Fig. 2. The sif view of a distributed system

sif's users are (typically) application developers—individuals responsible for developing applications to support data sharing; end-users are (typically) researchers that interact with applications built on top of sif.

The sif middleware exposes as much to the application developer as is considered useful for the application in question. It follows that it is the responsibility of the application developer to determine how much underlying detail is to be exposed to the end-users: it may be appropriate to expose the whole underlying data structure, allowing users to construct SQL queries; alternatively, a simple interface supporting pre-formulated queries might be appropriate—resulting in much less flexibility. A 'portal' approach—in which users can construct queries dynamically via a graphical interface—is also possible, and becoming increasingly popular; the benefits of such an approach are articulated in [22]:

> "For bench researchers, the web-services world might not look very different from the current one. Online databases would still exist, each with its own distinctive character and user interface. But bioinformaticists would now be able to troll the online databases to aggregate data simply and reliably. Furthermore, software engineers could create standard user interfaces that would work with any number of online data sources. This opens the door to genome 'portals' for those researchers who prefer to access multiple data sources from a single familiar environment."

The middleware is, essentially, a collection of web services. Quite deliberately, sif utilises only a small number of the core web services standards. The utilisation of standards and standards-based software has made the middleware easier to port to multiple operating systems: the middleware has been deployed successfully to machines running (various flavours of) Linux, Windows XP, IBM AIX, and Mac OS X.

sif offers support for three types of 'plug-in'—data plug-ins, file plug-ins and algorithm plug-ins—and it is these plug-ins that facilitate interoperability. By using a standard plug-in interface, it becomes possible to add heterogeneous resources into a virtual organisation; crucially, there is no need for the resource being advertised through the plug-in system to directly represent the physical resource—what is advertised as a single data source may represent any number of physical resources, or even another distributed system.

Data plug-ins treat all data sources as SQL databases, with the plug-in being responsible for all translations between the native data format and SQL. The plug-in user or application developer can retrieve schemas for known resources; as such, the user of the plug-in has the potential (depending, of course, on relevant permissions) to perform a query on data from fundamentally different data sources. From an end-user perspective, a user can request a list of plug-ins and descriptions from the middleware: definitions for a specific plug-in can be retrieved which contain an interface definition. The user can then request action to be taken for a plug-in by sending a request which conforms to the interface definition (the response is also bound by the definition so the response will be in an expected format).

sif acts primarily as a secure gateway and data integration framework. If, in a distributed context, a user runs a query across several data nodes, then the middleware will distribute that query to the nodes and aggregate the results. The reason that sif can expose any relational database is that it makes no assumptions about structure or semantics: while sif facilitates distributed queries, it is up to the end-user (or application) to ensure that the queries (and results) are

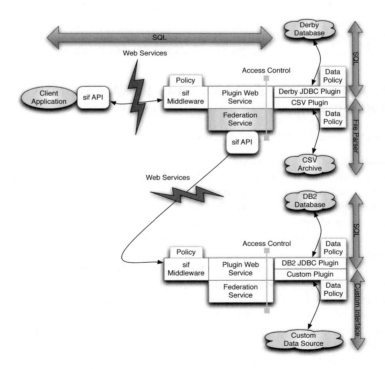

Fig. 3. sif: architecture

meaningful (recalling the classifications of Ouksel and Sheth from Section 1). This, of course, makes the task of federation much easier.

The architecture of a sif deployment is given in Figure 3. sif can be thought of as being comprised of three parts: the core middleware, the plug-ins, and the client-side API.

The core middleware manages the installed plug-ins, giving them a standard interface to be written against. It also provides a federation service to facilitate the construction of queries against multiple data sources. As all data is represented as if it were a standard SQL database, these queries take the form of SQL queries across distinct data sources each exposed via a separate plug-in. The access control framework enforces policies created by the owners of the data and the owners of the machine on which sif is being hosted. This allows data owners to restrict the data they expose to users, and server owners to control who the permitted users of services are.

The middleware has built-in capabilities for transferring files and data: installing, removing and updating plug-ins; advertising and defining resources exposed by plug-ins; and providing system status information. The core middleware exposes this functionality through a number of web services, all of which utilise strong cryptography to ensure privacy. The client-side API is a wrapper around web service calls to create the simplest possible interface for a new application developer to implement against; it also provides a number of helper functions to assist in common tasks.

Having described our middleware framework, we now illustrate how it has been used to support Systems Biology research within Oxford.

4   On the Sharing of Codes and Data to Support Cancer Modelling

4.1   Background: Cell Based Chaste

Mathematical and computational models of biological systems are rapidly increasing in complexity. This is especially true in fields such as cancer modelling, where the volume of available biological data is increasing exponentially. Modelling approaches therefore span the range from detailed models of molecular level changes and interactions that take place in the initial stages of cancer, right through to mechanical models of the mechanics of tumour development at the tissue level.

The objective of the Cell Based Chaste (Cancer, Heart and Soft Tissue Environment) (or *Cancer Chaste*) initiative [23,24,25] is to develop a mathematical and computational model that bridges across these spatial and temporal scales within a single, generic modelling framework. The focus of the project was initially on colorectal cancer, due to the wealth of particularly good experimental data that was available, and also because the biological understanding of the disease is sufficiently advanced to allow a systems-level approach.

Colorectal cancers originate from the epithelium that covers the luminal surface of the intestinal tract, with this epithelium renewing itself more rapidly than

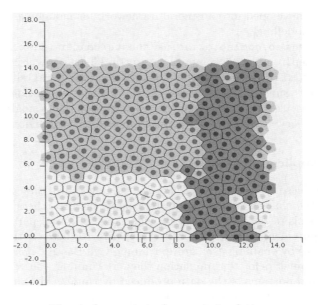

Fig. 4. A snapshot of a crypt simulation

any other tissue. This process of renewal requires a coordinated programme of cell proliferation, migration and differentiation, which begins in the tiny crypts of Lieberkühn that descend from the epithelium into the underlying connective tissue. It is generally believed that carcinogenesis occurs as a consequence of changes that disrupt normal crypt dynamics. Identifying the mechanisms that govern crypt dynamics is therefore essential to understanding the origins of colorectal cancer. The Cell Based Chaste team have developed a multiscale model of intestinal crypt dynamics that comprises three main components: models of intracellular signalling pathways; cell-cycle models; and a mechanical model that controls cell adhesion and migration at the macroscale. In recent years, the code

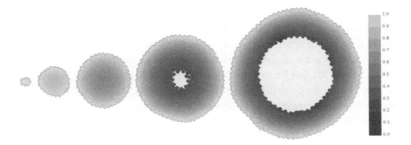

Fig. 5. Simulation of a growing multicell tumour spheroid, showing formation of a necrotic core

base has been developed into a general framework for modelling tumour growth and tissue remodelling.

Figure 4 (from `web.comlab.ox.ac.uk/chaste/cancer_index.html`) illustrates a single time step from a crypt simulation using the Chaste code; Figure 5 (also from `http://web.comlab.ox.ac.uk/chaste/cancer_index.html`) shows a number of time steps from a simulation of a growing tumour. Theories relating to stem cells in the crypt are discussed in [23] including the observation that a single cell's progeny can dominate a crypt, via a process termed monoclonal conversion.

4.2   The Application of sif

In combining these two activities (Cell Based Chaste and sif), we are undertaking to facilitate the sharing of simulation data between researchers with a view to enabling analysis to be performed on a wide range of initial parameters.

A key requirement in this respect is the desire to increase the co-operation between groups by enabling each group running Cell Based Chaste to store various initial parameters for the simulation they are running, the version of Chaste being used, and portions of the data produced by running the simulation—such as meshes and pointers to any animations generated. This will allow for the possibility of re-running the simulation at a later date with the same version of the software to generate any additional data that may be needed for performing additional analysis. The data will be held locally to each of the research groups and the sif middleware will be utilised to enable the groups to gain access to each others' data. Each group will, however, maintain access control over their own data and can choose when to release it to others. It is hoped that by sharing this data wider ranges of parameters can be investigated leading to more interesting results. Additionally, any analysis will be based on a larger data set—giving rise to greater confidence in results.

Our method for achieving this goal is illustrated in Figure 6.[5] The diagram shows researchers initially running the Cell Based Chaste code with a number of different starting parameters and for differing simulation times. The initial parameters are stored in a database along with the resulting simulation data and additional information such *as did the simulation turn monoclonal and if so after what period of time?*

After collecting the resulting data from a number of simulations the researchers can use the query application which utilises the federation features of the sif middleware to enable queries to be run across multiple data sets. This allows researchers to perform queries to identify the starting parameters that generate the type of resulting simulation they wish to investigate further. As an example, this may involve identifying the initial conditions necessary to produce monoclonal results and then investigate the surrounding parameter space, by running additional simulations, to identify which parameters have strongest affect on the tendency to become monoclonal. Such sharing of data also has the

---

[5] The shaded / dashed entities representing the notion of including additional nodes and data sets.

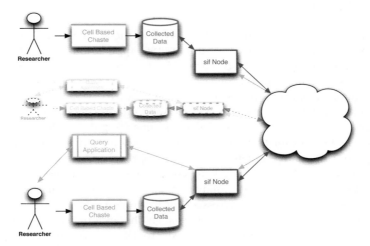

Fig. 6. Overview of data sharing

potential to help to reduce duplicated effort—allowing CPU time to be utilised more effectively by running simulations with new starting parameters, rather than having many people repeating the same simulation only to receive predictable results. This should give rise to a more coordinated process.

4.3   A Second Application

A second application being undertaken within OCISB, which also employs sif, is concerned with building histo-anatomically detailed individualized cardiac models [26]. To achieve this, the group requires access to large numbers of MRI images, which are typically stored on file systems accessible via protocols such as FTP and secure FTP.

We have utilised the sif middleware's plug-in architecture to provide a secure method for transferring these MRI images securely. The sif middleware's API provides a consistent interface to resources of the same type to application developers, enabling them to more readily develop applications with consistent user interfaces. Using sif, the data owner can easily control who has access to the data without having to create a new login account for each person—or have multiple people share a single password. Applications written using the middleware also enable users to access data from multiple sources without having to remember multiple passwords. The application that has been developed allows the user to select the data source of interest and list the available MRI collections that are accessible to the user—with access to resources being controlled by fine-grained policies; the user then selects the data set they are interested in, then the application retrieves the necessary data and processes it to produce a cardiac model.

5 Conclusions

In this paper we have provided a brief overview of the sif middleware framework, and outlined how it is being used to support two applications within the Oxford Centre for Integrative Systems Biology. The Centre has data integration and data sharing at the heart of its concerns; as such, we envisage sif supporting an increasing number of applications as we move forward.

Other work in the literature is concerned with integrating heterogeneous data within particular areas: [27] for SNP descriptions for pharmacogenomic studies and [28] for data relevant to Drosophila researchers are but two examples. Our focus, though, is the development of a generic approach. There are, perhaps, parallels to be drawn with B-Fabric [29], a "core framework for integrating different analytical technologies and data analysis tools". We have, though, consciously avoided issues of workflow, as addressed, for example by [30].

Much work remains to be done, not least in the area of schema transformations: while our approach to date has been to leave it to domain experts to resolve differences, we are fully aware that this philosophy will not serve us in the longer term. The work of [31] is of particular interest in this respect, as is that of [32].

It is, perhaps, instructive to reflect upon the differences between the healthcare research domain—which is, after all, the domain with we have the most familiarity and that for which sif was originally developed—and the Life Sciences. The most mundane difference is the size of images and data: the two applications that we are dealing with require files of significant size to be transported; this has required a minor redesign of the underlying code base. There are also differences in terms of security: in the Life Sciences, there is, inevitably, less of a concern with respect to privacy; however issues of confidentiality—in terms of protecting one's intellectual assets—are an important consideration. Finally (and here it should be stated that we are writing from the position of Computer Scientists who engage with collaborators from the relevant disciplines) there is one striking similarity across the two domains: a 'digital divide' would appear to be emerging between groups—those who are in a position to embrace emerging tools and technologies for effective data management will clearly prosper in the coming years, while those groups which are not so fortunate cannot possibly benefit. (This is a phenomenon which we have observed locally—within the OCISB—nationally, and globally.) As such, we would argue that issues of engagement and accessibility need to rise up the list of priorities of those groups who are developing technological solutions for data integration in the Life Sciences; to this end, this is one of our immediate concerns.

Acknowledgements

The authors acknowledge the support of the BBSRC—via the Oxford Centre for Integrative Systems Biology—in undertaking this work. We would also like to thank David Power, Douglas Russell, Tahir Mansoori and James Osborne for their valuable contributions.

References

1. Kim, J.: Phyl-OData (POD) from tree of life: Integration challenges from yellow slimy things to black crunchy stuff. In: Cohen-Boulakia, S., Tannen, V. (eds.) DILS 2007. LNCS (LNBI), vol. 4544, pp. 3–5. Springer, Heidelberg (2007)
2. Simpson, A.C., Power, D.J., Russell, D., Slaymaker, M.A., Kouadri-Mostefaoui, G., Ma, X., Wilson, G.: A healthcare-driven framework for facilitating the secure sharing of data across organisational boundaries. Studies in Health Technology and Informatics 138, 3–12 (2008)
3. Simpson, A.C., Power, D.J., Slaymaker, M.A., Politou, E.A.: GIMI: Generic infrastructure for medical informatics. In: Proceedings of the 18th IEEE Symposium on Computer-Based Medical Systems (CBMS 2005), pp. 564–566. IEEE Computer Society Press, Los Alamitos (2005)
4. Simpson, A.C., Power, D.J., Russell, D., Slaymaker, M.A., Bailey, V., Tromans, C.E., Brady, J.M., Tarassenko, L.: GIMI: the past, the present, and the future. To appear in the Philosophical Transactions of the Royal Society A (2010)
5. Tromans, C., Brady, J.M., Power, D.J., Slaymaker, M.A., Russell, D., Simpson, A.C.: The application of a service-oriented infrastructure to support medical research in mammography. In: Proceedings of MICCAI-Grid 2008 (2008)
6. Simpson, A.C., Slaymaker, M.A., Yap, M., Gale, A.G., Power, D.J., Russell, D.: On the utilisation of a service-oriented infrastructure to support radiologist training. In: Proceedings of the 22nd IEEE Symposium on Computer-Based Medical Systems (CBMS 2009). IEEE Computer Society Press, Los Alamitos (2009)
7. Ouksel, A.M., Sheth, A.: Semantic interoperability in global information systems. SIGMOD Record 28, 5–12 (1999)
8. Slaymaker, M.A., Power, D.J., Russell, D., Wilson, G., Simpson, A.C.: Accessing and aggregating legacy data sources for healthcare research, delivery and training. In: Proceedings of the 2008 ACM Symposium on Applied Computing (SAC 2008), pp. 1317–1324 (2008)
9. Slaymaker, M.A., Power, D.J., Russell, D., Simpson, A.C.: On the facilitation of fine-grained access to distributed healthcare data. In: Jonker, W., Petković, M. (eds.) SDM 2008. LNCS, vol. 5159, pp. 169–184. Springer, Heidelberg (2008)
10. Russell, D., Power, D.J., Slaymaker, M.A., Kouadri Mostefaoui, G., Ma, X., Simpson, A.C.: On the secure sharing of legacy data. In: Proceedings of the 2009 IEEE Conference on IT: Next Generation (ITNG 2009), pp. 1676–1679 (2009)
11. Stein, L.D.: Integrating biological databases. Nature Reviews Genetics 4(5), 337–345 (2003)
12. Lenzerini, M.: Data integration: a theoretical perspective. In: Proceedings of the twenty-first ACM SIGMOD-SIGACT-SIGART symposium on Principles of database systems, pp. 233–246 (2002)
13. Ives, Z.G.: Data integration and exchange for scientific collaboration. In: Paton, N.W., Missier, P., Hedeler, C. (eds.) DILS 2009. LNCS (LNBI), vol. 5647, pp. 1–4. Springer, Heidelberg (2009)
14. Paton, N.W.: Data integration in the life sciences: Fun, findings and frustrations. In: Bairoch, A., Cohen-Boulakia, S., Froidevaux, C. (eds.) DILS 2008. LNCS (LNBI), vol. 5109, pp. 8–10. Springer, Heidelberg (2008)
15. Ives, Z.G., Khandelwal, N., Kapur, A., Cakir, M.: ORCHESTRA: Rapid, collaborative sharing of dynamic data. In: Second Biennial Conference on Innovative Data Systems Research (CIDR 2005), pp. 107–118 (2005)

16. Kirsten, T., Rahm, E.: BioFuice: Mapping-based data integration in bioinformatics. In: Leser, U., Naumann, F., Eckman, B. (eds.) DILS 2006. LNCS (LNBI), vol. 4075, pp. 124–135. Springer, Heidelberg (2006)
17. Halevy, A.Y., Ives, Z.G., Mork, P., Tatarinov, I.: Piazza: data management infrastructure for semantic web applications. In: Proceedings of the 12th international conference on World Wide Web, pp. 556–567 (2003)
18. Ng, W.S., Ooi, B.C., Tan, K.L., Zhou, A.: PeerDB: A P2P-based system for distributed data sharing. In: Dayal, U., Ramamritham, K., Vijayaraman, T.M. (eds.) Proceedings of the 19th International Conference on Data Engineering, pp. 633–644. IEEE Computer Society press, Los Alamitos (2003)
19. Leser, U., Naumann, F.: (Almost) hands-off information integration for the life sciences. In: Second Biennial Conference on Innovative Data Systems Research (CIDR 2005), pp. 131–143 (2005)
20. Boeckmann, B., Bairoch, A., Apweiler, R., Blatter, M.C., Estreicher, A., Gasteiger, E., Martin, M.J., Michoud, K., O'Donovan, C., Phan, I., Pilbout, S., Schneider, M.: The SWISS-PROT protein knowledgebase and its supplement TrEMBL in 2003. Nucleic Acids Research 31(1), 365–370 (2003)
21. Power, D.J., Politou, E.A., Slaymaker, M.A., Simpson, A.C.: Towards secure grid-enabled healthcare. Software: Practice and Experience 35(9), 857–871 (2005)
22. Stein, L.D.: Creating a bioinformatics nation. Nature 417, 119–120 (2002)
23. Pitt-Francis, J., Pathmanathan, P., Bernabeu, M.O., Bordas, R., Cooper, J., Fletcher, A.G., Mirams, G.R., Murray, P., Osborne, J.M., Walter, A., Chapman, S.J., Garny, A., van Leeuwen, I.M.M., Maini, P.K., Rodriguez, B., Waters, S.L., Whiteley, J.P., Byrne, H.M., Gavaghan, D.J.: Chaste: a test-driven approach to software development for biological modelling. Comp. Phys. Comm. 180, 2452–2471 (2009)
24. van Leeuwen, I.M.M., Mirams, G.R., Walter, A., Fletcher, A., Murray, P., Osborne, J., Varma, S., Young, S.J., Cooper, J., Pitt-Francis, J., Momtahan, L., Pathmanathan, P., Whiteley, J.P., Chapman, S.J., Gavaghan, D.J., Jensen, O.E., King, J.R., Maini, P.K., Waters, S.L., Byrne, H.M.: An integrative computational model for intestinal tissue renewal. Cell Proliferation 42, 617–636 (2009)
25. Pathmanathan, P., Cooper, J., Fletcher, A., Mirams, G., Murray, P., Osborne, J., Pitt-Francis, J., Walter, A., Chapman, S.J.: A computational study of discrete mechanical tissue models. Physical Biology 6(3), 36001 (2009)
26. Plank, G., Burton, R.A., Hales, P., Bishop, M., Mansoori, T., Bernabeu, M.O., Garny, A., Prassl, A.J., Bollensdorff, C., Mason, F., Mahmood, F., Rodriguez, B., Grau, V., Schneider, J.E., Gavaghan, D., Kohl, P.: Generation of histo-anatomically representative models of the individual heart: tools and application. Philosophical Transactions of the Royal Society A: Mathematical, Physical and Engineering Sciences 367(1896), 2257–2292 (2009)
27. Coulet, A., Smaïl-Tabbone, M., Benlian, P., Napoli, A., Devignes, M.-D.: SNP-converter: An ontology-based solution to reconcile heterogeneous SNP descriptions for pharmacogenomic studies. In: Leser, U., Naumann, F., Eckman, B. (eds.) DILS 2006. LNCS (LNBI), vol. 4075, pp. 82–93. Springer, Heidelberg (2006)
28. Zhao, J., Miles, A., Klyne, G., Shotton, D.: OpenFlyData: The way to go for biological data integration. In: Paton, N.W., Missier, P., Hedeler, C. (eds.) DILS 2009. LNCS (LNBI), vol. 5647, pp. 47–54. Springer, Heidelberg (2009)

29. Türker, C., Stolte, E., Joho, D., Schlapbach, R.: B-fabric: A data and application integration framework for life sciences research. In: Cohen-Boulakia, S., Tannen, V. (eds.) DILS 2007. LNCS (LNBI), vol. 4544, pp. 37–47. Springer, Heidelberg (2007)

30. Pettifer, S., Wolstencroft, K., Alper, P., Attwood, T., Coletta, A., Goble, C., Li, P., McDermott, P., Marsh, J., Oinn, T., Sinnott, J., Thorne, D.: [my]Grid and UTOPIA: An integrated approach to enacting and visualising in silico experiments in the life sciences. In: Cohen-Boulakia, S., Tannen, V. (eds.) DILS 2007. LNCS (LNBI), vol. 4544, pp. 59–70. Springer, Heidelberg (2007)

31. Zamboulis, L., Martin, N., Poulovassilis, A.: Bioinformatics service reconciliation by heterogeneous schema transformation. In: Cohen-Boulakia, S., Tannen, V. (eds.) DILS 2007. LNCS (LNBI), vol. 4544, pp. 89–104. Springer, Heidelberg (2007)

32. Bowers, S., Ludäscher, B.: An ontology-driven framework for data transformation in scientific workflows. In: Rahm, E. (ed.) DILS 2004. LNCS (LNBI), vol. 2994, pp. 1–16. Springer, Heidelberg (2004)

Helping Biologists Effectively Build Workflows, without Programming

Paul M.K. Gordon[1,2], Ken Barker[1], and Christoph W. Sensen[2]

[1] Department of Computer Science, University of Calgary, 2500 University Dr. NW, Calgary, Canada T2N 1N4
[2] Department of Biochemistry and Molecular Biology, University of Calgary, 3330 Hospital Dr. NW, Calgary, Canada T2N 4N1
{gordonp,kbarker,csensen}@ucalgary.ca

Abstract. Seahawk is a browser for Moby Web services, which are online tools using a shared semantic registry and data formats. To make a wider array of tools available within Seahawk, the Daggoo system helps users adapt forms on existing Web sites to Moby's specifications. Biologists were interviewed and given workflow design tasks, which revealed the types of tools present in their conceptual analysis workflows, and the types of control flow they understood. These observations were used to enhance Seahawk so that Moby and external Web tools can be browsed to create workflows "by demonstration". A flow-up user study measured how effectively biologists could 1) demonstrate a workflow for a realistic task, 2) understand the automatically generated workflow, and 3) use the workflow in the Taverna workflow editor/enactor. The results show promise that biologists without programming experience can become self-sufficient in analysis automation, using workflow-by-demonstration as a first step.

Keywords: programming by demonstration, bioinformatics workflows, user study, semantic web services.

1 Introduction

There is much anecdotal evidence that biological researchers manually perform repetitive analyses. However, empirical data regarding workflow needs and adoption amongst biologists is lacking. Interview and survey studies of software use by bioinformatics end-users has mostly been done by groups in Library Sciences attempting to understand the issues in order to build support mechanisms [1,2]. Within the bioinformatics community itself, most research has been to understand types of tasks and classify them for ontological use [3,4,5], or to develop novel interaction methods [6]. The widespread use of spreadsheets environments such as Excel to organize and process data using cell formulae is evidence that learning some end-user programming is seen as worthwhile by biologists as their datasets grow. Unfortunately, spreadsheets cannot perform many of the analyses that could be coordinated using a full programming language. For most biologists, the leap in functionality afforded by

P. Lambrix and G. Kemp (Eds.): DILS 2010, LNBI 6254, pp. 74–89, 2010.

a traditional programming language is more than offset by the difficulty in learning it. Visual workflow languages (discussed in Section 2.1) try to find a balance between advanced functionality and usability by programming novices.

Clearly, learning any workflow programming environment will still put some cognitive burden on the biologist. Is it possible to make the learning curve even gentler? Blackwell argues [7] that programming is best understood as activities involving both loss of direct manipulation and introduction of notation. It is possible to avoid programming (visual or otherwise) altogether using a technique known as programming-by-demonstration. More sophisticated than macro recording, the semantics of the users' direct manipulations are understood by the system and translated into programming language notation on their behalf. In workflow-by-demonstration (WbD) systems, a user interactively calls a service (an individual program or on-line tool) and uses the output as input to another service, building up a chain of actions. Visiting services f, g, and h in that order is interpreted as a workflow program equivalent to $h(g(f(x)))$. This model has been used in the past by Seahawk [8], Gbrowse Moby [9], and Galaxy [10] to generate workflows automatically.

This paper describes three activities: a user study to inform design, an implementation of WbD, and an evaluation of the implementation. Through formal studies of and interviews with biologists, this paper finds evidence that for WbD to be practical, it must be able to capture logic more complex than a linear sequence of service calls. The implementation of additional features is detailed, as well as evaluated for practicality via a small-scale user study using biologists.

2 Bioinformatics Workflows

An exhaustive review of existing workflow systems and programming-by-demonstration systems is beyond the scope of this work. Instead, original observations drawn from several small-scale user studies will highlight barriers to successful workflow creation by a wider audience of biologists.

2.1 Biologists' Reality

To date there has only been a small amount of research focused on software engineering issues for professional programmers in bioinformatics [6], and virtually none on bioinformatics end-users. Despite this lack of empirical evaluations, there is no shortage of tools purporting to simplify analysis tasks for users. For professional bioinformatics programmers there are large open-source code libraries for the Perl [11] and Java [12] languages that have been built by the developer community over the years, but these are beyond the scope of casual end-user programmers.

More esoteric professional programmer environments in the field of bioinformatics, PROVA [13] and Opal [14] bear the closest resemblance to domain-specific rule systems. Opal uses XML syntax to define the interface to an application, and can use the Kepler [15] visual workflow environment to create automated analysis. An early example of a bioinformatics visual workflow construction tool was TAMBIS [16], and it is widely cited inside and outside the bioinformatics community for introducing an ontology-driven workflow model. TAMBIS was an early predecessor to the most well-establish bioinformatics workflow tool currently in use, Taverna [17]. A pilot Taverna user study [18] determined that a major usability issue is data input types/formatting/connecting. Taverna is fundamentally data format agnostic, allowing users to easily create I/O mismatches.

Seahawk generates Taverna workflows using an ontology-driven service model (Moby [19]). A larger audience of Life Science researchers may adopt workflow programming as cognitively simpler, type-safe environments such as Seahawk reach a wider audience, and do not require them to explicitly conceptualize the analysis programs being created. Other prominent existing tools for bioinformatics analysis automation all take somewhat different interface approaches and have different activity building blocks. In all cases, the workflow must first be constructed and then tested with sample data, thereby forcing the user to conceptualize the program explicitly. Some examples are:

- Remora [20]: resembles Taverna insofar as a workflow is constructed visually from service components, but workflow construction is type-safe and stepwise via a Web interface. Build blocks: Moby Web Services.
- BioPipe [21]: uses a hand-written XML specification file for the workflow. Building blocks: a set of Perl scripts interfacing to BioPerl.
- Integrator [22]: uses XQuery for service mediation, plus a procedural instruction set to integrate data from multiple Web resources using Collection and Filtering steps. Building blocks: XML formatted on-line data.
- Wildfire [23]: has a GUI workflow builder with advanced control elements such as 'for each' loops, and concentrates on grid-based computing of large datasets. Building blocks: programs accessed via ACD-defined interfaces.
- Bio-STEER [24]: has a relatively simple GUI and composition mechanism. Building blocks: Web services described semantically in OWL-S.

Of particular relevance to the work presented here, the Bio-STEER authors assert:

Recognizing the potential benefit of Semantic Web technologies...the use of semantics to filter and suggest only semantically compatible services will reduce the time it takes to construct a meaningful workflow.

Type-safe Moby services are at the core of Seahawk for this reason. Clear semantic descriptions of the data being manipulated, and services used in a workflow demonstration, are essential to correctly capture the biologist's intentions in WbD.

2.2   Biologists' Needs and Intuitions

To date there has been no close examination of how biologists naturally think about workflows. Software usability can be improved by providing a close mapping between users' natural tendencies and interface implementation. In a new study, 16 participants who had never used Taverna were briefly introduced to the concept of workflows. The participants were all molecular biologists, including graduate students, professors and industrial researchers. All of them were familiar with at least popular resources such as the NCBI Web site, BLAST, and the Gene Ontology (GO).

On average, participants reported using on-line bioinformatics tools more than once a week. The average input data list was 5-10 items long.

The participants were paired off and given 30 minutes to invent and produce "pen and paper" workflows that they thought could help them in their research. They discussed the workflows with their partner to promote workflow clarity. All of the participants grasped the concept of a workflow (based on written feedback), and twelve of the biologists (75%) devised a workflow that could help them in their own research.

The workflows were analyzed along three dimensions that correspond to major features of workflows: data, functions, and control structures. Input data to the workflows was categorized as either actual data, or database identifiers. Seven different types of input data were present in the workflows, 2 types of database identifiers, and 2 pseudo-database identifiers (namely gene symbols and organism names). This split suggests that both information retrieval and information processing workflows are important to biologists. Genome DNA sequences and chromatograms (raw sequencing machine data files) were the only two input data types that appeared more than once, showing a diversity of inquiry starting points amongst the study subjects.

An additional type of data identified was a custom database to be searched. Three workflows contained BLAST searches against real or hypothetical private datasets. Although these types of searches are important, because the vast majority of activities (51 in total) used in the workflows have Web-based implementations, local services will not be examined further in the current work.

The activities or "functions" performed in the workflows were categorized in a way analogous to data versus identifiers: manipulation activities versus retrieval activities. Unlike the input data, activities were clearly lopsided towards manipulation (44) rather than retrieval (7). This indicates that bioinformatics workflows are not merely a substitution for a distributed query language, but rather they are procedural automation systems. In any sufficiently powerful procedural system, it is important to have flow control mechanisms such as conditional blocks and loops. Of the workflows examined, 3 had "if" conditions and 4 had "if/else" type constructs. Noteworthy is that in all four "if/else" instances, the else was logically represented as the negation of the if condition (i.e. $if(x)\{...\}if(!x)\{...\}$). This and other observations informed the design of Seahawk's data filters, discussed in Section 3.2.

With regards to looping control, a distinction can be made: loops over input data versus loops over intermediate results. The higher the number of loop iterations, the higher the automation payoff is for the biologist. In every workflow, the input was a list of data or identifiers that could range between tens and thousands of items in length. The utility of iterating over input lists was obvious to the 75% of study participants who devised workflows for their research. Less obvious to the biologists, though still relevant, is the work saved by automatically looping over intermediate results. Four of the workflows contained substantial list "explosions", where one input data leads to many results that need subsequent processing before a final workflow output is produced.

Another pertinent observation from the workflows was that two included wet-lab activities in the middle of the workflow, such as cloning, assays, amplification and sequencing. Interviews with biologists were conducted to reveal barriers to analysis automation. Issues related to the correctness of their programs (e.g. Excel formulae) were frequently brought up by interviewees, especially if the data is to be used as the basis for further experimentation:

"...if it had irreversibly changed [my cloning] construct, and I didn't notice, then I'll be working in the lab with bad data and not really knowing it"

When questioned about the decision to do some processes manually the most common response was pressure to meet impending deadlines and although it took more effort to do analysis manually, it ensured correct results because of the certainty that transformations were being done correctly. If they were to learn how to program the analysis, they may not have enough time to program *and* test the correctness. Another interviewee describes the advantage of direct manipulation in the analysis:

"I had to think about stepwise what the flow would be for the data, and it was just a matter of like 'okay, try this out. Okay, this step. Okay, organize it well. Okay, now that it's organized I [can figure out]... the next step."

These observations suggest that a demonstrational approach, rather than a direct programming approach, can lower barriers to automating analysis.

The interviews also revealed another contributing factor to the continued manual use of multiple, separate pieces of analysis software. Despite the availability of integrated packages for many tasks, researchers want to follow the "gold standard" analysis for their research area of interest. The analysis tools and methods used in the most cited papers end up being the preferred ones to use, whereas using newer, integrated solutions has the potential to raise questions with reviewers unfamiliar with the latest tools. This need to perform very particular analyses (typically meaning the use of a particular set of Web sites) is a key motivation for the user-driven integration of arbitrary Web forms into Seahawk workflows, described in Section 3.1.

In summary, the key points drawn from the user study are that: 1) biologists desire workflows involving both data and identifiers, 2) conditions and iteration are essential even for novices, 3) workflow construction should be step-wise, with intermediate results shown, and 4) users must be able to automate arbitrary Web forms in the workflows. These observations were used to inform the design of Seahawk's WbD system.

2.3 Biologists' Expectations

Observation of biologists' use of earlier versions of Seahawk lead to several salient observations about their expectations and tendencies during the demonstration phase of WbD. Initially, users were instructed to manipulate one sample piece of data, and told that their actions would be generalized for batch input processing in a workflow later. Despite this instruction, many users attempted to highlight swathes of data, hoping to process all of them interactively and simultaneously. Explicit 'for each' looping for each intermediate step was expected by users, and was therefore implemented in the Seahawk service menus (Fig. 1). Importantly, rather than a promise of efficiency in workflow creation, the 'for each' feature demonstrates immediate efficiency versus manual analysis, without additional user effort.

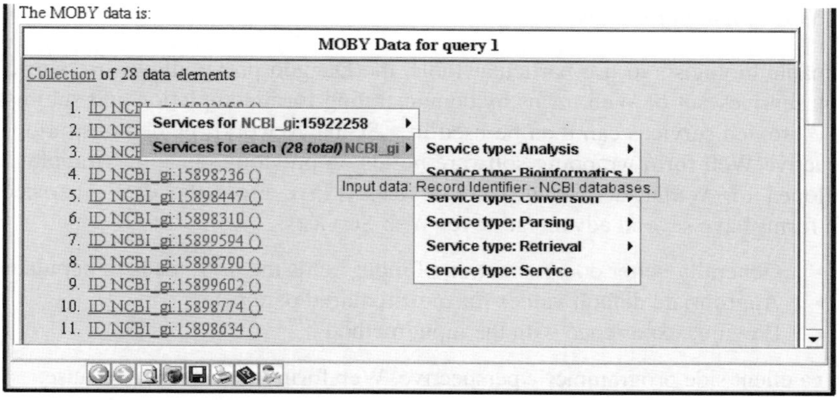

Fig. 1. Seahawk UI showing option to run service for each collection member

The collective display of intermediate results for input lists is also important for the creation of conditionals, because having both positive and negative example data helps the biologist verify that they have applied the correct filter (see Section 3.2). Immediate dataset feedback can help mitigate the fear of programming errors mentioned by several of the interviewees. One can think of 'for each' over peer lists as somewhat like cell range selection in spreadsheets but more flexible because the cardinality of the lists does not need to be predetermined.

An observation particular to the Seahawk interface was that users did not necessarily read the top-level service menu choices because they were focused on looking at service names in submenus. Since the top-level menu choice defines the semantics of the data, ignoring the labels could lead to incorrect assumptions by Seahawk when generating the workflow. A WbD system must know *why* an action is taken, but the user is concentrated on *what* action must be taken and can easily overlook the semantic labels of the interface. In this case, simply highlighting the data type and namespace information in red helped users remain mindful of the data semantics they were implying with their menu choice.

The final major observation accounted for in Seahawk is that data used as input to a workflow can vary greatly in original formatting. Since Taverna is largely data-format agnostic, a user's inability to input data correctly to the generated workflow can be a serious barrier to success (first documented elsewhere [18]). Alleviating manual reformatting is a prime motivator for the aides described in Section 3.3.

3 Implementation

While Seahawk has multiple UI elements, examined here are those newly developed behaviours that were inspired by the study-derived insights into biologists' workflow needs. Detailed videos demonstrations of semantic Web form wrapping, data flow control, and workflow preview/enactment are available at http://www.daggoo.net.

3.1 Web Form Semantics

To enable biologists to use particular tools, the Daggoo proxy allows users to create Moby services out of Web forms by demonstrating form completion and submission. These proxied services can then be used in a Seahawk workflow demonstration. The interactive Web form wrapping software builds on previous semantic wrapping code developed for WSDL-described Web services ([25]). From the user's perspective, Web forms have several advantages over Web Services:

- Generally better documentation of input fields the user needs to populate
- Appropriate default values for most required parameters
- Previous experience with the input method

From a client side programmer's perspective, Web forms are difficult because:

- Web forms are not natively supported in Taverna
- The form may rely on dynamic HTML display or submission using AJAX
- Output parsing of HTML is more difficult than Web Service XML parsing

Seahawk tries to address each of these concerns. Proxying the Web forms as Moby Services both adds semantics to the service and allows Taverna to enact them. The interactive completion and submission of the form is done in the user's Web browser (Fig. 2), mitigating many problems due to AJAX dependencies. With regards to results parsing; Seahawk's database of regular expression rules helps discover data, regardless of its HTML tag nesting, with the tradeoff of possible mis-selection if the output format is highly variable.

The semantic wrapping process consists of 5 steps:

1. Dragging the Web page containing the form *onto* Seahawk.
2. Filling in the form in the browser, using data dragged *from* Seahawk.
3. Submitting the Web form.
4. Selecting in Seahawk the data to return, from a list of automatically recognized data types.
5. Filling in metadata for the new Moby service, e.g. service name, free text description, service type.

This procedure captures as much of the service specification as possible via demonstration. Input parameters and their default values are automatically gleaned from the form's HTML, while input semantic types are determined by tracking the Seahawk origin of fields populated via drag and drop. The wrapping proxy captures the submission method whether JavaScript or HTML based. The possible semantic types of the returned data are enumerated for the user, using a database of XPath and regular expression rules, whose format was described earlier [25]. Manual metadata specification allows the biologist to capture the meaning of the service beyond the formalism of data type signatures. This will help other biologists to decide if they can reuse the service in their own workflow (wrapped services are added to the public Moby service registry).

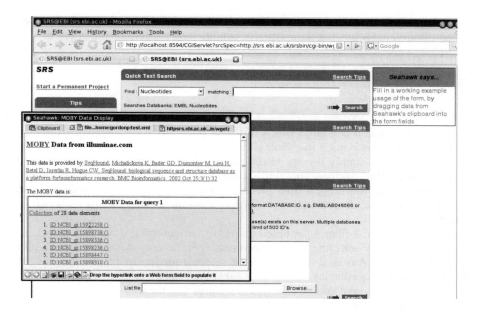

Fig. 2. Daggoo Web form in user's browser (with form fill in tips from Seahawk)

The Daggoo proxy server, which includes management features for wrapped services, has been designed to run on Google's cloud computing service to avoid the Moby service proxy becoming a bottleneck in workflows. Current restrictions on port availability and timeouts prevent complex analysis services from being deployed in the cloud but these restrictions should ease in the future. Another limitation is that some popular bioinformatics sites, such as the NCBI Web portal, rely on AJAX in the response. Currently, Daggoo does not evaluate AJAX responses, but users can similarly wrap NCBI Web Services in Daggoo instead.

The demonstrational nature of the wrapping process has two main advantages. First, it provides a test case that can be run to ensure the proxy is still functioning properly in the face of Web site changes. Second, the sample data used in the Web form is part of a Seahawk history, which allows just-in-time wrapping of services to seamlessly be part of a workflow demonstration.

3.2 Dataflow Control

Seahawk implements conditional execution via a novel search/filter widget. Since every UI action must be mapped to a workflow component in some way for WbD, the behaviour of the search/filter must be carefully considered. The behaviour is a hybrid of a highlighting search and a gray-out filter. Filtering is essential so that the user cannot interact with data not meeting the search criteria: these data will not be available downstream in the equivalent workflow. To avoid such an inconsistent state, filtering is updated as each letter is typed into the search box.

Balancing the needs of the workflows examined, and the need for interface comprehensibility, the following three types of conditionals are currently supported:

1. $if(cond(x))\{f(x)\}$
    $else\{g(x)\}$
2. $if(cond(x.member1))\{f(x.member2)\}$
    $else\{g(x.member2)\}$
3. $if(cond(f(x)))\{g(x)\}$

Currently, the search condition can be either a plain string or a regular expression. Building on the observation that biologists thought of *else* conditions as $if(!cond(x))$ so *else* conditions are simply inverse selections based on switching the keyword "if" to "unless" in the filter widget. This fulfils conditionals of type 1 listed above. In terms of the equivalent workflow, a filter processor with two output ports forks the workflow (Fig. 3). In the demonstration, forking is accomplished by calling a service in a new tab (Shift+click).

Conditional type 2 filtering is based on particular members of the data objects and is accomplished by selecting from a document-specific drop-down list at the end of the search phrase. Any data member can be selected for further processing using the hyperlinks in the data display, unless they are grayed-out by the filter (Fig. 4).

The demonstration of conditional type 3 requires that the user reference data item x from the output of $f(x)$. In Seahawk, this is implemented by providing a 'for previous input' service menu (Fig. 4) when a filter is applied to a service result (i.e. $f(x)$). If no filter is applied to a service result, the 'previous input' option is not displayed because

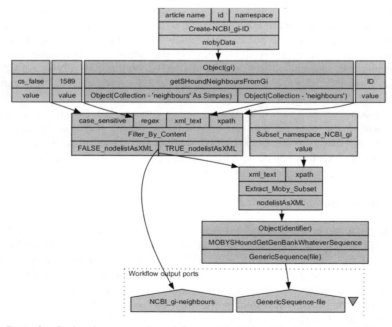

Fig. 3. Part of a Seahawk generated workflow, with if/else filtering represented by the two output ports of the *Filter_By_Content* processor

Fig. 4. UI for 'previous input' option, and equivalent workflow processors. *MOBYSHoundGet-GenBankWhateverSequence* can be called on the original GI number (15922258) because it has neighbours (from service *getSHoundNeighboursFromGi*) passing the filter "ID contains 1589".

this translates into $g(f(x))$, which is the default interpretation of the service browsing history. The equivalent set of Taverna workflow activities requires non-trivial list manipulation processors, and would almost certainly be beyond the capabilities of a workflow programming novice.

Other somewhat unusual aspects of the search/filter behaviour are page-specificity and job-level filtering. Page specificity means that the filter widget is tied to the given service output, so when a user navigates away from a page, the widget disappears. Upon reentry to the given page it will reappear. Hiding the search/filter widget disabled the filtering, re-showing the widget re-enables the previous filter criteria. This stateful, page specific behaviour is essential to allowing different conditions to be applied at different stages of the workflow. Statefulness also means that output from one services can be filtered multiple ways by launching subsequent services in new tabs after various search edits (a variation on the if/else concept).

Job level filtering reflects the for-each functionality in Seahawk. In a document containing results for 10 inputs (i.e. 10 'jobs'), the filtering decision is to either allow or disallow the subsequent processing of each job individually: they are independent results. Having lists of jobs helps filter development because users receive immediate feedback on the correctness of positive and negative examples, as opposed to surmising the correctness of the filter for arbitrary data based on a single data sample.

The multiple novel features introduced in the filter/search widget have the potential to be barriers to correct workflow demonstration, and therefore are one of the primary targets of the user study conducted below.

3.3 Workflow Visualization and Enactment

Building the workflow from the browsing history gets the biologist much of the way to automating analysis, but the workflow must also be practical. Keeping in mind that a workflow is essentially a set of processors with data links, Seahawk takes into account practical considerations to make them maximally useful to the biologist:

- Preview the workflow, for correctness
- Parameterize processors such as conditionals, for modifiability

- Minimize processor redundancy, for efficiency
- Name processors with semantic information, for comprehensibility
- Maximize workflow metadata, for readability
- Include parsing processors, for usability

Before workflow export, the user is given the opportunity to preview the generated workflow (Fig. 5). The workflow image is currently generated by an automated call to a remote Ruby CGI script. This allows them to quickly go back and adjust their Seahawk browsing, filtering, *etc.* if the workflow does not capture their intent. Once they are happy with their workflow, it can be saved, and loaded in Taverna.

Keeping in mind that scientific analysis can be somewhat trial-and-error, making it easy to modify parameters to behaviours such as filtering and decomposition, even once in Taverna, could be very useful to the biologist. To this end, behaviour parameters are clearly marked as external constants in Seahawk workflows, although this can crowd the display. Minimizing processor redundancy can help alleviate this crowding, and involves Seahawk tracking: 1) when an input has been used more than once, 2) when a filter is applied multiple times, 3) when the same object member is extracted for further processing again, and 4) filter negations.

Both the number of processors and their labeling is important. Since the data and services are semantically typed, Seahawk can generate meaningful labels for otherwise anonymous nodes in a workflow, such as inputs and outputs. Example values gleaned from the demonstration are also added as metadata to the workflow.

By providing fields for additional metadata in the workflow preview such as a title and free text description of the whole analysis, Seahawk promotes readability, and reusability of the generated workflows. The actual enactment of the workflow is outside the control of Seahawk because the workflow is run by the Taverna GUI. One

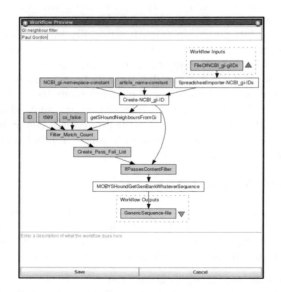

Fig. 5. Seahawk's workflow export preview, with fields for metadata addition by the biologist

barrier that Seahawk attempts to focus in on data input where, for example, a spreadsheet importer in the workflow allows users to specify their input lists within the familiar environment of rows and columns in Excel.

4   User Study

The purpose of conducting a user study was to evaluate the:

1. usability of Seahawk for demonstrating workflows (Can they do what they want?),
2. comprehensibility of the generated workflows (Do they know what they did?), and
3. viability of the generated workflows (Can they automate what they did?)

Implicit in points 1 and 2 is the question of whether or not the generated workflows are a faithful representation of the workflows they intended. Point 3 is critical to determining the practicality of WbD as a real solution to user-driven data integration.

The practicality of workflow programming for the biologist is based on the answer to the question: Is there a net benefit to learning how to build computerized workflows? Blackwell's Attention Investment Model [7] identifies key factors users consider when deciding to learn programming:

- *Cost:* how long would it take to do it manually?
- *Investment:* what is the time required to learn how to program it?
- *Pay-off:* how much time does the automation save?
- *Risk:* what is the probability of failure to write the correct program?

A WbD system should minimize risk and investment, while maximizing pay-off. The cost of the equivalent manual analysis must also be determined. The study subjects were asked to perform two exercises:

1. Given 9 GI numbers for nucleotide (EST) sequences, retrieve the sequences. If the sequence contains the promoter motif, design a hybridization probe using the Web-based Primer 3. Output each sequence containing the motif and its probe.
2. Given a GO term (seed growth, GO:0080112), find the related genes in the Arabidopsis genome and BLAST the sequences to find the equivalent rice gene.

Half of the users did the first exercise manually and the second exercise using Seahawk. The other half of the users undertook the tasks in the reverse order. These workflows were based on the example workflows described in Section 2.2, with both incorporating a conditional and a fork. The first task includes an input list, while the second contains an intermediate list expansion. The tasks also include both existing Moby services and wrapping a Web form (the *Primer3* Web form and the *arabidopsis.org* GO search, respectively). Both loop over 9 data items. Approximately 10 data elements were used because this reflects the average input list size reported by researchers in another study (manuscript submitted). When the task was performed in Seahawk, the workflow was also exported and executed in Taverna. Users were asked to qualitatively describe their comprehension of the workflow and their overall perception of using Seahawk to help automate the task.

5   Results

The initial phase of this study had eight participants, all of whom are molecular biology graduate students. Each session lasted approximately 1.5 to 2 hours. Each task was therefore performed four times in Seahawk and four times manually. The remaining time was split between an introductory video to Seahawk, task description, loading Taverna, and a post-task interview. Task completion times were analyzed using a single factor ANOVA F-test to determine if manual and Seahawk-based analyses truly differ in duration (Table 1).

Table 1. Time to complete bioinformatics tasks, with ANOVA significance

	Manual ($\mu \pm \sigma$)	Seahawk ($\mu \pm \sigma$)	F-test
Task 1 *in minutes*	27.5±5.8	18.75±5.5	4.79 (p = 0.071)
Task 2 *in minutes*	33.75±6.6	22.75±3.5	8.47 (p = 0.027)

A major qualitative observation of the study is that the users did not appear to have difficulty with the novel filter/search mechanism. Furthermore, with one exception, all users gravitated naturally towards using the for-each menu items. The major barrier timewise in Seahawk was in the metadata specification and parameter selection during Web form wrapping. For Task #2, the p-value of 0.027 is below the widely accepted 0.05 threshold for statisical significance. This implies that completing the task in Seahawk rather than manually is indeed faster, even for novices.

6   Discussion

6.1   Validity

The modest sample size for the initial evaluation means that the results should be regarded as preliminary, but promising, and justify further study. Typically, threats to the validity of a study are based on inappropriate metrics and conceptualization (construct), inconsistency of experiment execution (internal), or inapplicability to the problem being modeled (external). Blackwell's Attention Investment Model, used for the study's metrics and observations is highly cited, supporting construct validity.

All participants were given tasks from the same task set and Seahawk software version. The same introductory video was shown to all with no additional assistance in using the interface. All of the Moby services and Web forms were available during all of the sessions, with negligible differences in response times. The internal validity of the experiment was therefore not greatly threatened.

The programming tasks used in the experiment are drawn from actual workflows described by biologists so the tasks were both non-trivial and realistic. This supports external validity of the experiment.

How useful to the larger population of Life Scientists is WbD? Key to the approach presented here is 1) the availability of on-line resources to coordinate, 2) common ontologies/semantics for the data types, and 3) recognition of database identifiers (and sequences where available). Genomics and proteomics databases used by molecular biologists are already intricately cross-linked with biomedical data, which additionally has its own ontologies for disease, taxonomy, *etc*. With regards to sequence recognition, Seahawk contains rules to recognize common formats for the third type of biopolymer, glycans, for which research tools are becoming increasing available [26]. Metabolomics should also be amenable to workflows, which has seen a growth in online resources that could be wrapped, such as the Human Metabolome Database [27]. Further afield, ecology informatics has rich semantics [28], but its formalization is nascent and there are relatively few systematic on-line resources.

6.2 Demonstration Techniques

The Daggoo Moby service proxy was motivated by the need to let researchers use the online resources with which they were already familiar. Our preliminary results show that users are quite capable of semantically wrapping these services for inclusion in workflows. The extra time required to do the wrapping (*investment*) versus calling the service manually *(cost)* is recouped before 9 iterations of a 'for each' loop (*payoff*). This payoff would be even more pronounced for larger loops because the wrapping step is a fixed rather than incremental cost. This makes it easier to convince biologists to adopt Seahawk if an appropriate pipeline for their analysis does not already exist.

Workflow by demonstration in Seahawk is no different than just using Seahawk interactively for a task. In follow up interviews, the test subjects perceived little *risk* in performing the analysis in Seahawk rather than manually in a Web browser. The primary issue raised was one of interactivity: at times users thought Seahawk was stalled when it was actually calling services or creating menus and dialog boxes. This highlights the importance of constant feedback, which made WbD an attractive option in the first place. Note that the perception of risk is as important as actual risk because the risk-based decision is made before even attempting to use a new tool.

6.3 Workflow Usability

All of the participants successfully saved their workflows and launched them in Taverna, which was preloaded on the computer. Somewhat surprisingly, none of the participants found the workflow graphics particularly confusing. Intimate familiarity with the task just performed likely raises the level of comprehension, and is a subject for further exploration with regards to workflow sharing and reuse. Two users took several minutes to learn how to load the input data file but all successfully enacted the workflow. Overall, the participants reported seeing value in a workflow based approach, particularly for publication of their work.

7 Conclusions

This research has found that workflows could be useful for most biologists, who often need to process lists of data. Despite this and the plethora of workflow tools available, researchers usually run such analyses manually. This is due to several factors including fear of failure in programming and the inability to use a particular Web site in the workflow. Seahawk addresses the former, Daggoo the latter.

Key to providing a powerful WbD system is having strong semantics for the data, services, and UI actions such as search/filter and hyperlink based object decomposition. Preliminary results indicate that with appropriate sample data, researchers can build practical, complex workflows with this vocabulary of direct manipulations. Some open questions remain in WbD. For example, how can Seahawk workflows merge multiple service results on the clipboard, while still allowing users to delete clipboard items? Currently, data coming from the clipboard is treated as top-level workflow inputs. Defining the semantics of clipboard item deletion and other Seahawk actions will allow novice users to create even more complex workflows without any programming.

Further structured user testing of the Seahawk/Daggoo WbD system will be conducted, in the hopes of identifying remaining barriers to workflow programming, as perceived by biologists. Bioinformatics software empirical evaluation should help software engineers by providing more powerful and more approachable automation tools. Getting more biologists started on the learning curve towards self-sufficiency will undoubtedly pay dividends in terms of increased productivity.

References

1. MacMullen, W., Vaughan, K., Moore, M.: Planning bioinformatics education and information services in an academic health sciences library. Coll. Res. Libr. 65, 320–332 (2004)
2. Rein, D.C.: Developing library bioinformatics services in context: the Purdue University Libraries bioinformationist program. J. Med. Libr. Assoc. 94, 314–320, E193-E197 (2006)
3. Stevens, R., et al.: A classification of tasks in bioinformatics. Bioinf. 17, 180–188 (2001)
4. Tran, D., Dubay, C., Gorman, P., Hersh, W.: Applying task analysis to describe and facilitate bioinformatics tasks. Medinfo. 11, 818–822 (2004)
5. Bartlett, J.C., Toms, E.G.: Developing a protocol for bioinformatics analysis: An integrated information behaviour and task analysis approach. J. Am. Soc. Inf. Sci. 56, 469–482 (2005)
6. Kulyk, O., Wassink, I.H.C.: Getting to know Bioinformaticians: Results of an exploratory user study. In: Proc. BCS HCI 2006 Int'l Workshop on Combining Visualisation and Interaction to Facilitate Scientific Exploration and Discovery, pp. 30–37 (2006)
7. Blackwell, A.F.: First steps in programming: a rationale for attention investment models. In: Proc. IEEE 2002 Symp. on Human Centric Computing Lang. and Env., pp. 2–10 (2002)
8. Gordon, P.M.K., Sensen, C.W.: Seahawk: Moving Beyond HTML in Web-based Bioinformatics Analysis. BMC Bioinformatics 8, 208 (2007)
9. Wilkinson, M.: GBrowse Moby: a Web-based browser for BioMoby Services. SCBM 1, 4 (2006)
10. Giardine, B., et al.: Galaxy: a platform for interactive large-scale genome analysis. Genome Res. 15(10), 1451–1455 (2005)
11. Stajich, J.E., et al.: The BioPerl toolkit: Perl modules for the life sciences. Genome Research 12, 1611–1618 (2002)
12. Pocock, M., Down, T., Hubbard, T.: BioJava: open source components for bioinformatics. SIGBIO Newsletter 20, 10–12 (2000)
13. Kozlenkov, A., Schroeder, M.: Prova: Rule-based java-scripting for a bioinformatics semantic web. In: Rahm, E. (ed.) DILS 2004. LNCS (LNBI), vol. 2994, pp. 17–30. Springer, Heidelberg (2004)

14. Krishnan, S., et al.: Opal: Simple Web Services Wrappers for Scientific Applications. In: Proc. IEEE Conf. Web Services, pp. 823–832 (2006)
15. Ludaescher, B., et al.: Scientific workflow management and the Kepler system. Concurr. Comput.: Pract. Exper. 18, 1039–1065 (2006)
16. Stevens, R., et al.: Tambis: transparent access to multiple bioinformatics information sources. Bioinformatics 16(2), 184–185 (2000)
17. Oinn, T., et al.: Taverna: A tool for the composition and enactment of bioinformatics workflows. Bioinf. 20, 3045–3054 (2004)
18. Gordon, P.M.K., Sensen, C.W.: A Pilot Study into the Usability of a Scientific Workflow Construction Tool. Tech. Report 2007-874-26, University of Calgary (2007)
19. Wilkinson, M.D., et al.: Interoperability with Moby 1.0: It's Better than Sharing Your Toothbrush! Briefings in Bioinformatics 9(3), 220–231 (2008)
20. Carrere, S., Gouzy, J.: Remora: Pilot in the ocean of Biomoby Web Services. Bioinf. 22, 900–901 (2006)
21. Hoon, S., et al.: Biopipe: A Flexible Framework for Protocol-Based Bioinformatics Analysis. Genome Res. 13(8), 1904–1915 (2003)
22. Chagoyen, M., Kurul, M.E., De-Alarcon, P.A., Carazo, J.M., Gupta, A.: Designing and executing scientific workflows with a programmable integrator. Bioinformatics 20(13), 2092–2100 (2004)
23. Tang, F., et al.: Wildfire: distributed, grid-enabled workflow construction and execution. BMC Bioinf. 6, 69 (2005)
24. Lee, S., Wang, T.D., Hashmi, N., Cummings, M.P.: Bio-STEER: A semantic web workflow tool for grid computing in the life sciences. Future Gener. Comput. Syst. 23(3), 497–509 (2007)
25. Gordon, P.M.K., et al.: Using a novel data transformation technique to provide the EMBOSS software suite as semantic web services. In: Proc. IEEE Int'l Conf. Bioinf. Biomed., pp. 117–124 (2007)
26. Turnbull, J.E., Field, R.A.: Emerging glycomics technologies. Nat. Chem. Biol. 3, 74–77 (2007)
27. Wishart, D., et al.: HMDB: a knowledgebase for the human metabolome. NAR 37, D603 (2009)
28. Michener, W.: Meta-information concepts for ecological data management. Eco. Inf. 1, 3–7 (2006)

A Data Warehouse Approach to Semantic Integration of Pseudomonas Data

Kamar Marrakchi[1], Abdelaali Briache[1], Amine Kerzazi[2], Ismael Navas-Delgado[2],
José Francisco Aldana-Montes[2], Mohamed Ettayebi[3], Khalid Lairini[1],
and Badr Din Rossi Hassani[1]

[1] Department of Biology, Faculty of Sciences and Techniques, BP: 416 - University
Abdelmalek Essaâdi, Tangier, Morocco
[2] Department of Computer Languages and Computing Science, Higher Technical School of
Computer Science Engineering University of Málaga, 29071, Malaga, Spain
[3] Department of Biology, Faculty of Sciences Dhar-Mahraz, Sidi Med Ben Abdellah
University, B.P. 1796 Atlas, Fez, morocco
{mkamar,a_briache,kerzazi,ismael,jfam}@lcc.uma.es,
{moettayebi,bdrossi}@gmail.com,
Khalidlairini@yahoo.com

Abstract. Biological research and development are routinely producing terabytes of data that need to be organized, queried and reduced to useful scientific
knowledge. Even though data integration can provide solutions to such biological problems, it is often problematic due to the sources' heterogeneity and their
semantic and structural diversity. Moreover, necessary updates of both structure
and content of databases provide further challenges for an integration process.
We present a new biological data warehouse for Pseudomonas species "*PseudomonasDW*" to integrate annotation and pathway data from highly different
resources. The combination of knowledge from multiple disciplines and sources
should advance the understanding of cellular processes and lead to the prediction of cellular behavior in its entirety. The key aspect of our approach is the
combination of a materialized and a virtual data integration to exploit their advantages in a new hybrid approach. The data are extracted from the original
data sources using SB-KOM (System Biology Khaos Ontology-based Mediator) and then stored locally in the data warehouse to ensure a fast performance
and data consistency.

Keywords: Data Integration, Data Warehouse, Web Services, Ontology, Pseudomonas.

1 Introduction

Biological experimentation or analysis often involves the generation of large quantities
of data. These data are frequently most useful in conjunction with other biological data,
so managing and integrating such data collected by scientists and researchers from all
around the world can be very valuable. This is an exceedingly challenging problem for
a wide variety of reasons. These reasons include the quantity of biological data, the

P. Lambrix and G. Kemp (Eds.): DILS 2010, LNBI 6254, pp. 90–105, 2010.
© Springer-Verlag Berlin Heidelberg 2010

large number of biological databases, the rapid rate in the growth of biological data, the overabundance of data types and formats, the variety of data access techniques, database heterogeneity and errors in biological data; in addition to the interdisciplinary nature of bioinformatics.

Today, one of the big challenges of bioinformatics is to allow the biologists to efficiently reach several data sources, by means of a unified global schema, via automatic procedures [1]. This automation should succeed to a real cooperation between biologists and machines, for efficient research and better exploitation of the results. Also, it has been extensively demonstrated that an integrative approach is effective to discover some relationships between these data. When all the data produced or published are integrated in a singular system, biological research will be conducted in manner that is quite unlike the way it was done before. Researchers will be able to process massive amounts of complex data much more quickly and they will garner insight about the areas of their interest rapidly. In view of this, many efforts have been led to share genomics and biological data for their use in integration projects [2].

1.1 Data integration Approaches

The two main technologies used to resolve the problem of data integration are the mediator-based integration and data warehouse systems:

- Mediator approach basically translates user queries into queries that are understood by the sources integrated into the system. This approach maps the relationship between source descriptions and the mediator and thus allows queries on the mediator to be rewritten as queries on the data source. The two main approaches for establishing the mapping between each source schema and the global schema are global as view (GAV) and local as view (LAV) [3,4].
- A data warehouse is an approach to physically integrate all the relevant data in a central data store. Warehouse integration consists of materializing the data from multiple sources into a local warehouse and executing all the queries on the data contained in the warehouse rather than in the sources. Warehousing emphasizes on data translation, as opposed to query translation in mediator-based integration [5]. In fact, data warehousing requires that all the data loaded from the sources be converted through data mappings to a standard unique format before it is physically stored locally. The major function of data warehouse is to load and translate data from different external sources into one large database. For this process, imported data are carefully mapped using ETL-processes (Extraction, transformation and loading) [25]. Thereafter, all the information in the data warehouse is quickly accessible.

1.2 Motivation

There are various benefits and drawbacks in these two approaches, as discussed in the previous paragraph. Querying of data warehouses is simplified since queries can be performed against a single data warehouse which contains all of the integrated information. However, the creation of a data warehouse can be a very difficult, large-scale process that includes global data integration involving many types of heterogeneity. Moreover, data warehouse design is difficult since the data warehouse should probably be optimized for the various types of queries. Another limitation to using a warehouse is that, since a warehouse is updated periodically from its constituent databases, data can become out-of-date.

In this paper we introduce *PseudomonasDW*, a semi structured data warehouse for integrating the heterogeneous biological information collected from highly different external resources including KEGG [14], BRENDA [15], UniProt [16], GenBank [17] and PRODORIC [18]. The combination of knowledge from multiple disciplines and data sources should advance the understanding of cellular processes and will lead to the prediction of cellular behavior in its entirety. In our system we combine a materialized and a virtual data integration to exploit their advantages in a new hybrid environment. Our goals were (i) the inclusion of available high-throughput genomics data, (ii) the integration of various data sources using the combinatorial approach outlined and (iii) keep data up-to-date.

PseudomonasDW focuses on *Pseudomonas* species. We have chosen this well-studied bacterium because of its importance in the biomedical and the agricultural domains. In fact, it is considered a valuable control agent for some plant illnesses [6]. Pseudomonas species play a major role in the medical and the agricultural domains. With *PseudomonasDW* we aim to help biologist to understand and explain the biological processes of interest by using an integrative system.

This paper is organized as follows: an overview on some biological data warehouse is given in the next section. In section 3, the provenance and the format of structured resources used for integration in *PseudomonasDW* system are described. In section 4, the integration process along with some explanatory schemas is presented. The data mapping rules that have been defined for instances conciliation during the integration process are also presented. Section 5 sketches the *PseudomonasDW* content. Section 6 concludes the paper.

2 Related Works

In bioinformatics, data warehouses can be roughly separated into two groups: general software infrastructures for further customization within new bioinformatics applications (e.g., ONDEX [7], Altas [8], BioWarehouse [9], BioMart [10]) and project-oriented data warehouse implementations for particular biological questions (e.g., Systomonas [11], Columba [12]).

ONDEX is a system for automated ontology alignment, ontology-based text indexing and database integration. It focuses on recently developed features for graph-based analysis and visualization. This system was applied to the task of interpreting gene expression results and was also used to discover functional annotations for most of the genes that emerged as significant in the microarray experiment, but were previously of unknown function [7].

The Atlas biological data warehouse serves as data infrastructure for bioinformatics research and development. Atlas achieves integration of diverse data sets at two levels. First, Atlas stores data of similar types using common data models, enforcing the relationships between data types. Second, integration is achieved through a combination of APIs, ontology and tools [8].

BioWarehouse is an open-source toolkit for constructing bioinformatics databases using different database management systems. It facilitates and allows several database integration tasks like comparative analysis and data mining. It integrates its component databases into a representational framework and enables multi-database querying using SQL [9].

BioMart is a simple and robust data integration system for large scale data querying. It has been designed to provide researchers with an easy and interactive access to both the wealth of data available on the Internet and for in house data integration [10].

Systomonas provides an integrated bioinformatics platform for a systems biology approach and the biology of pseudomonas in infection and biotechnology. Apart from the in-house experimental metabolome, proteome and transcriptome data, it also stores the prediction information of cellular processes such as gene regulatory networks [11].

Columba is an integrated database of proteins, structures and annotations and is specially designed to support the creation of sets of protein structures sharing annotations in any of the data sources [12].

Both systems, Systomonas and Columba, are available via a web-based graphical user interface that can be used with any web browser.

3   PseudomonasDW Overview

PseudomonasDW is a semi structured data warehouse which was conceived with the goal of meeting biologists' expectations in terms of the *Pseudomonas* genomics, proteomics and metabolic data. We find in the literature [13] the distinction of two approaches in the construction of data warehouses respectively named procedural and declarative approaches. In the procedural approach the data are integrated without trying to construct an integrator schema. In this case, we talk more often about the Data Repository than the data warehouse, whereas in the declarative approach, the data are structured taking advantage of its global schema or integrator schema. *PseudomonasDW* is based on the declarative approach, even though this approach is more complex but more often consistent. The model in which the global schema is defined determines the query language used to interrogate the warehouse.

3.1   PseudomonasDW System Architecture

PseudomonasDW consists of several components that contribute to the data integration process in different ways. These single software components and their collective interaction enable the realization of *PseudomonasDW*. The two main technologies applied in our system are data warehouse and mediator-based integration. We combine a materialized and a virtual data integration to exploit their advantages in a hybrid environment. On the one hand, the data warehouse offers high performance for complex data. On the other hand, up-to-date data can be retrieved when needed.

A schematic representation of *PseudomonasDW* system architecture is shown in Figure 1. The source layer is the basis of the system and contains the data access to the sources KEGG [14], BRENDA [15], UniProt [16], GenBank [17] and PRODORIC [18].

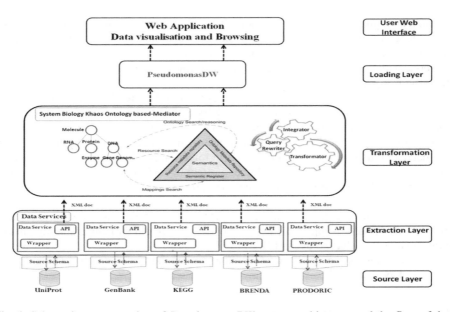

Fig. 1. Schematic representation of *PseudomonasDW* system architecture and the flow of data from the original heterogeneous data sources to the data warehouse.

In *PseudomonasDW* system, the public data sources are uniformly accessed and integrated through SB-KOM (System Biology Khaos Ontology-based Mediator) [19] which offers wrapper interfaces to data sources and translates data into common data model used by the mediator. Thus, these wrappers provide the Extraction Layer. In SB-KOM, XML is used at the source level (in wide sense XML, XML schema and XQuery as the common data model). It uses ontologies as integration schemas, and takes advantage of simple reasoning to perform the query rewriting (see section 4.3), enabling the Transformation and Loading Layers.

In this proposal the ETL process is started by the administrator by selecting the species that will be stored in *PseudomonasDW*. Then the system extracts all the wished and the available data for these species via these wrappers, transforms the extracted data into unique form XML and stores them in the data warehouse using the different compounds of SB-KOM. Finally, the system is automatically updated when changes are detected in the sources, by means of the incremental approach.

Our proposal is to use ontology to integrate the data, where each source is related with the global schema by means of mappings. A detailed description for the different system components is cited in the next section.

3.2   Data Resources

Several resources could be used to create a data warehouse like *PseudomonasDW*. We describe data sources that have been selected for having the most appropriate properties for studying pseudomonas species: 1) genomic and proteomic databases, 2) metabolic databases and 3) enzymatic resources. We will show in this section for each selected resource, its provenance, content and structure.

3.2.1 Genomic and Proteomic Databases

PseudomonasDW provides various genomic data such as gene and protein annotation, gene regulation, gene expression and a collection of transcription factor binding sites data. These data are extracted from three most commonly used databases:

- **GenBank:** is a comprehensive database that contains publicly available nucleotide sequences for more than 300 000 organisms. It is accessible through the NCBI Entrez retrieval system, which integrates data from the major DNA and protein sequence databases along with taxonomy, genome, mapping, protein structure and domain information, and the biomedical journal literature via PubMed. GenBank is one of the first banks that proposed XML format for its records with a well-defined DTD specifying the structure and the domain terminology for the records of genes and submitted sequences [17].
- **UniProt:** is a comprehensive resource for protein sequence and annotation data that provides also information concerning protein function, and contains 516603 sequence entries, comprising 181919312 amino acids sequences abstracted from 188500 references[1]. It provides data in HTML, XML and Fasta format [16].
- **PRODORIC:** is an acronym for **PRO**cariot**IC** **D**atabase **O**f Gene-**R**egulation. This database is an integrated approach to provide information about molecular networks in prokaryotes with focus on pathogenic organisms. It provides a web service for accessing important parts of the database [18].

3.2.2 Metabolic Databases

The metabolic information is extracted from **KEGG** [14] which is an integrated database resource consisting of 16 main databases. In *PseudomonasDW* we are interested only in one of them, **KEGG PATHWAY** providing pathways maps for metabolism and other cellular processes, as well as human diseases; manually created from published materials. The users can access the KEGG API server by the SOAP technology over the HTTP protocol. The SOAP server also comes with the WSDL, which makes it easy to build a client library for a specific computer language. This enables the users to write their own programs for many different purposes and to automate the procedure of accessing the KEGG API server and retrieving the results.

3.2.3 Enzymatic Resource

PseudomonasDW provides enzymatic data extracted from **BRENDA** [15] which represents the largest freely available information system containing a huge amount of biochemical and molecular information on all classified enzymes as well as software tools for querying the database and calculating molecular properties. The database covers information on classification and nomenclature, reaction and specificity, functional parameters, occurrence, enzyme structure and stability, mutants and enzyme engineering, preparation and isolation, the application of enzymes, and ligand-related data.

---

[1] Release 2010_05 of 20-Apr-2010.

4 Bio-data Integration

Our primary goal is to integrate all these databases in a singular data warehouse. Integration of data from heterogeneous knowledge sources represents the consolidation of heterogeneous data geared at generating new knowledge that can not be obtained from single data sources.

In general, there are two levels of data integration in data warehouses. The first one is a syntactic integration which consists of extracting data from the original data sources and translating them into the uniform model used by *PseudomonasDW*. The second level is a semantic integration which consists of converting the extracted data in terms of the global schema by creating a correspondence between each source schema and the one of the warehouse.

4.1 Syntactic Data Integration

In *PseudmonasDW* system, we have chosen XML, in wide sense: XML, XML schema and XQuery, as the common data model. As a first step in building *PseudomonasDW*, we have created an XML schema for each data sources (see an example in Figure 2). These schemas are considered as models to define the different elements and attributes that describe the organization of the information within the original data sources.

Developing web services is a preliminary step in the SB-KOM architecture. These Web services are implemented in java, receive queries in XQuery and return XML documents. The goal of web services is to allow applications to access wrapper functionalities [20].

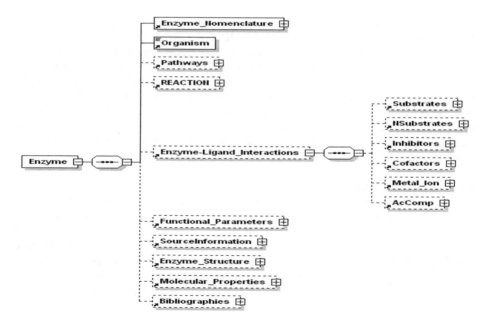

Fig. 2. A part of BRENDA Schema defined in *PseudomonasDW* system

In this context wrappers are an important part of the internal element of web services. A wrapper is an interface which receives query, accesses a data source, extracts data and translates them into a common data model used by SB-KOM (see Figure 3).

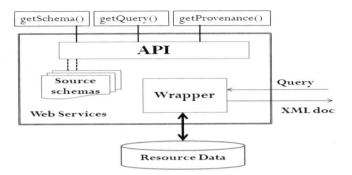

Fig. 3. Architecture of the web Services in *PseudomonasDW* system

The web services not only encapsulate a wrapper's query service, but also provide access to the schema of the information they store. For this purpose the *getSchema()*, *Query()*, and *getProvenance()* methods are published as an API (Application Programming Interface), and they return respectively the data's schema (see Figure 2), the XML document which must conform to its source schema and the data resource name. The later is a requirement in bioinformatics where it is important to know which information resource is being used (due to different data quality and reliability). Finally, all the methods provided by the web service were well-defined in a WSDL (Web Service Description Language) document [26].

4.2 Semantic Data Integration

One of the important components in a data warehouse system is its global schema or integrator schema. In the design of a data integration system, an important aspect is the way in which the global schema is specified, i.e., which data model is adopted and what kind of constraints on the data can be expressed. In our system we have followed the GAV approach (global-as view), which requires that the global schema be expressed in terms of the data sources.

Our proposal is to use an ontology (*PseudomonasDW* ontology) [27] to integrate data. *PseudomonasDW* Ontology (see Figure 4) provides a formal representation of the *Pseudomonas* domain by defining concepts and relationships between them. More precisely, to every concept of *PseudomonasDW* ontology, a view over the data sources is associated, so that its meaning is specified in terms of the data residing at the sources.

PseudomonasDW ontology contains 110 classes, 79 datatype properties that link individuals to data values and 44 object properties that link individuals to each other.

The process of registering data services for being integrated implies finding a set of mappings between *PseudomonasDW* ontology and data services schema. These

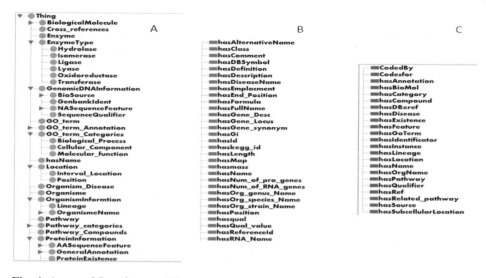

Fig. 4. A part of *PseudomoansDW* ontology. (A) presents some concepts in the ontology, (B) presents some Datatype properties and (C) presents some Object properties.

mappings will be key elements to integrate all the data sources, and will be used to make the resource semantics explicit. Moreover, the mappings we have used are defined as pairs *(P,Q)*. *P* is a set of path expressions on the resource schema (expressed by XPath) and *Q* a query expression in terms of the ontology (presented by conjunctive queries) [24]. To further explain relations shown in Figure 5, we give this example of mappings. In fact, there are three kinds of mappings:

Class Mapping:
> *Rule*: XPath Location, Class Local Name, Similarity-Value
```
/Result/Enzyme,Enzyme,100
/Result/Functional_Parameter,Functional_Parameter,100
```

Datatype Property Mapping:
> *Rule*: Domain XPath Location; XPath Location, Domain Class Local Name; Property Local Name, Similarity-Value
```
/Result/Functional_Parameter/KMs;/Result/Functional_Par
ameter/KMs,KMs;hasValue;100
/Result/Functional_Parameter/pH_OPTIMUM;/Result/Functio
nal_Parameter/pH_OPTIMUM,pH_OPTIMUM;hasValue;100
```

Object Property Mapping:
> *Rule*: Domain XPath Location; Range XPath Location, Domain Class Local Name; Range Class Local Name; Property Local Name, Similarity-Value
```
/Result/Enzyme;/Result/Functional_Parameter,Enzyme;Func
tional_Parameter;hasFunctionalPara,100
```

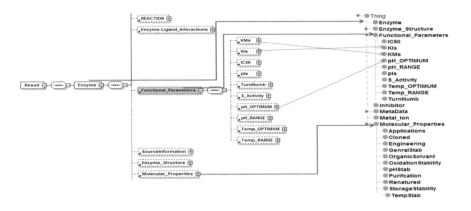

Fig. 5. Example of the mapping between a BRENDA schema and PseudomonasDW ontology. Source schema (left). PseudomonasDW ontology (right).

4.3 SB-KOM: System Biology Khaos Ontology-Based Mediator

To populate *PseudomonasDW* we have used SB-KOM mediator [19] which is based on KOMF [21]. KOMF is a generic infrastructure to register and manage ontologies, their relationships and also information relating to the resources. This infrastructure is based on a resource directory, called Semantic Directory [22, 23], with information about web resource semantics. KOMF has been successfully instantiated in the context of molecular biology for integrating biological data sources which are accessible via internet pages. SB-KOM mediator is composed of three main components (Figure 6):

- *The Controller*: which receives queries from *PseudomonasDW* and coordinates the mediator components in order to evaluate these queries and return the results.
- *The Query Planner*: the role of this component is to find a query plan (*QP*) for the queries submitted by *PseudomonasDW*. Plans generated by this component specify the data sources from which the information must be retrieved and in which order they must be accessed.
- *The Evaluator/Integrator*: this component analyzes the query plan (QP), and performs the corresponding calls to the data services involved in subqueries (SQ1,...,SQn) of the query plan. So this component executes the web services in an order specified by the query plan.

The controller creates different threads for different requests sent by *PseudomonasDW* administrator, and assumes the role of the middleware between the mediator components. Queries are expressed as conjunctive predicates [24], with three main type of predicate: classes in terms of *PseudomonasDW* like Protein(P), have just one argument (P in this case), datatype properties that link individuals to data values like ProteinName(P,"Value") that links a protein with its name, and object properties that link individuals to individuals like isEnzyme(P,E). Properties have two arguments: domain and range. The results of these queries are instances of the ontology which the query was expressed in.

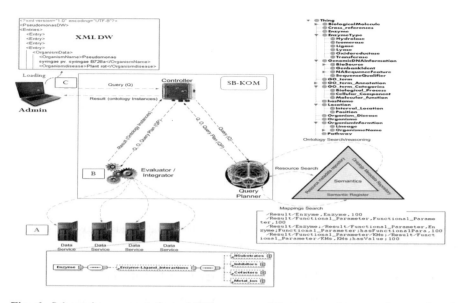

Fig. 6. Schematic representation of ETL process: (A) presents the step of extracting data, (B) presents the step of translating data and (C) presents the step of loading data in *PseudomonasDW*.

Once the conjunctive query is received at the level of the *Controller*, the request is sent to the *Query Planner* to elaborate one or several query plans to solve the conjunctive query from different data sources. Plans generated by the *Query Planner* component specify the data sources from which the information can be retrieved and in which order they must be accessed. The evaluation of these queries depends on the query plans themselves and must be generated expeditiously.

According to the query language (Conjunctive Queries), there will be different types of mapping in the semantic directory. Classes will be connected to the XPath of one or several XML Schema resource elements. Datatype properties will be connected to those two expressions: the first one corresponds to the class and the second to the property. The object properties will be related to the active XPath classes in the property. A possible conjunctive query sent by *PseudomonasDW* administrator is:

```
"Ans(P,E,O,G,PW):-Protein(P),hasName(P,"ProteinName"),
ForOrganism(P,O),Enzyme(E),IsEnzyme(P,E),Organism(O),hasO
rganismName(O,"OrganismName"),ForOrganism(E,O),Gene(G),
CodedBY(P,G),PathWay(PW), ParticipateIn(P,PW);"
```

This query is an example of a conjunctive query aiming to get information about a protein having name "ProteinName" (e.g *Acetyl-coenzyme A carboxylase carboxyl transferase subunit alpha*) for an organism called "OrganismName"(*Pseudomonas fluorescens (strain Pf-5)*). This query involves five databases: BRENDA, KEGG, UniProt, GenBank and PRODORIC.

In the query example, we have five classes (Protein, Organism, Enzyme, Gene and PathWay), two datatype properties (hasProteinName and hasOrganismName) and four object properties (ForOrganism, IsEnzyme, CodedBY and ParticipateIn).

This query will return instances of class Protein whose name is "ProteinName", and are related to:

- • The "Organism" by means of the relation ForOrganism.
- • The "Pathway" by means of the relation ParticipateIn.
- • The "Gene" by means of the relation CodedBY.
- • The "Enzyme" by means of the relation IsEnzyme. This Enzyme has to be related to the same Organism which the protein is related to by means of the relation ForOrganism.

A possible plan of execution for our query is shown in the Figure 7. The arcs represent object properties: IsEnzyme, ParticipateIn, CodedBY and ForOrganism in this case.

The plan contains all the information needed by the *Evaluator/Integrator* to execute a local query; this information includes resource URI and the local query expressed in XQuery. After that, these queries are executed at the level of the web services, the wrappers extract data from the original sources and return XML documents to the *Integrator* component. This one builds the instances (RDF model) from the data service results by using the mappings. These instances are not interconnected because they have been produced by different data services.

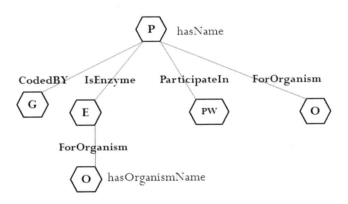

Fig. 7. Query plan for the example query described previously. Each node and arc contains information to access the corresponding data service.

In order to associate the different instances, the Integrator component establishes the relationships between services in the query plan and object properties defined in the ontology. Finally, these interrelated instances are filtered in order to eliminate the information not required. The final result is a set of *PseudomonasDW* ontology instances that includes all data extracted from the integrated data sources. These final instances are automatically transformed to XML document using a java API (e.g Jena) and loaded into an XML native data warehouse. Figure 6 illustrates the whole ETL process in *PseudomonasDW* system from the data services till Pseudomonas Data Warehouse.

5 Results

With SB-KOM, we have integrated data from five external sources, such as the information on transcription factor binding sites and data about gene regulation and gene expression are extracted from PRODORIC, the Metabolic reactions or pathways from BRENDA and KEGG and gene or protein annotation from GenBank and UniProt. The combination of knowledge from multiple disciplines and sources will drive the understanding of cellular process and lead to the prediction of cellular behavior in its entirety. Consequently, *PseudomonasDW* provides data from all levels of analysis as proteomics data, metabolite measurements, sequence data, gene-regulatory networks and corresponding enzyme data. The UML class diagram shown in Figure 8 permitted us to define the main concepts of *PseudomonasDW* and the relations that bind them as well as to understand the *PseudomonasDW* structure.

Currently, *PseudomonasDW* contains information on 20 different *Pseudomonas* species and strains. It contains more than 100000 proteins, 1000 enzymes, 50 regulons and 1780 pathways with their description.

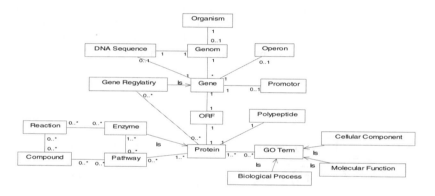

Fig. 8. *PseudomonasDW* UML conceptual Diagram

UniProt and GenBank have created mailing lists which are intended for distribution of messages announcing updates. Subscribing to these lists allows us to receive the latest modification and to keep track of updates to individual entries. PRODORIC, BRENDA and KEGG are updated periodically and provide comprehensive archives which contain only the updated entries. These archives allow us to specify which entries integrated in *PseudomonasDW* have been updated. Thus, Up-to-date annotation data is virtually integrated using a mediator, when the system is notified by the modified entries.

PseudomonasDW and SB-KOM are coupled by means of mediator queries. When the system receives an update notification from the administrator, the data services extract only the updated data from the original entries. Afterwards, it is possible to automatically start the integration process for updating the data warehouse by replacing only the obsolete data by the updated ones.

6 Conclusion

In this paper we have presented a semi-structured data warehouse for storing, managing and integrating the biological information collected from different data sources via the Web. *PseudomonasDW* focus on data integration of the biotechnologically and medically relevant group of bacteria, the *Pseudomonas* species.

In our system, we have combined a materialized and a virtual data integration to exploit their advantages in new hybrid environment. On one hand, data are physically stored in the data warehouse to be directly and quickly accessed. On the other hand, integrating and up-to-dating data is virtually achieved using mediator and is retrieved on demand according users' needs.

Wrappers are an important part in ETL process, where a wrapper is considered as an interface which receives query, accesses to a data source, extracts data and translates them into a common data model used by SB-KOM. The different components of the mediator (Controller, Query Planner and Evaluator/Integrator) are in charge of data translation step using schema integrator and mappings. As schema integrator is ontology (*PseudmonasDW* ontology) and results are ontology instances. We have developed *PseudmonasDW* ontology which provides a formal representation of the real world by defining concepts and relationships between them. The use of the ontology and instances enable reasoning to be included at different levels. The different instances returned by SB-KOM are loaded within *PseudomonasDW* after their automatic translation to XML format.

With *PseudomonasDW* we would like to provide biologists with accessible tools to elucidate cellular process of interest using an integrated systems strategy.

Acknowledgments. This work has been founded by the project PCI-EUROMED A/016116/08, P07-TIC-02978 (Junta de Andalucía), and TIN2008-04844 (Spanish Ministry of Science and Innovation).

References

1. Buttler, D., Coleman, M., Critchlow, T., Fileto, R., Han, W., Pu, C., Rocco, D., Xiong, L.: Querying multiple bioinformatics information sources: can Semantic web research help. ACM SIGMOD 31, 59–64 (2002)
2. The gene Ontology Consortium: The Gene Ontology (GO) database and informatics resource. Nucleic Acids Research 32, 258–261 (2004)
3. Florescu, D., Levy, A.Y., Mendelzon, A.O.: Database Techniques for the World-Wide Web: A Survey. ACM SIGMOD 27, 59–74 (1998)
4. Lenzerini, M.: Data Integration: A Theoretical Perspective. In: ACM Symposium on Principles of Database Systems (2002)
5. Sujansky, W.: Heterogeneous Database Integration in Biomedecine. Methodological Review. Journal of Biomedical Informatics 34, 285–298 (2001)
6. Allaire, M.: Diversité fonctionnelle des Pseudomonas producteurs d'antibiotiques dans les rhizosphères de conifères en pépinières et en milieu naturel. PhD Thesis, Faculty of Agriculture and Food University Laval Québec (2005)

7. Kohler, J., Baumbach, J., Taubert, J., Specht, M., Skusa, A., Rueegg, A., Rawlings, C., Verier, P., Philippi, S.: Graph-based analysis and visualization of experimental results with Ondex. Bioinformatics 22, 1383–1390 (2006)

8. Shah, S.P., Huang, Y., Xu, T., Yuen, M.M.S., Ling, J., Ouellette, B.F.F.: Atlas–a data warehouse for integrative bioinformatics. BMC Bioinformatics 6 (2005)

9. Lee, T.J., Pouliot, Y., Wagner, V., Gupta, P., Stringer-Calvert, D.W.J., Tenenbaum, J.D.: BioWarehouse: a bioinformatics database warehouse toolkit. BMC Bioinformatics 7 (2006)

10. Töpel, T., Hofestädt, R., Scheible, D., Trefz, F.: RAMEDIS: the rare metabolic diseases database. Applied Bioinformatics 5, 115–118 (2006)

11. Choi, C.C., Munch, R., Leupold, S., Klein, J., Siegel, I., Thielen, B., Benkert, B., Kucklick, M., Schobert, M., Barthelmes, J., Ebeling, C., Haddad, I., Scheer, M., Grote, A., Hiller, K., Bunk, B., Schreiber, K., Retter, I., Schomburg, D., Jahn, D.: SYSTOMONAS-an integrated database for systems biology analysis of Pseudomonas. Nucleic Acids Research 35, 537–537 (2007)

12. Trißl, S., Rother, K., Rother, Muller, H., Steinke, T., Koch, I., Preissner, R., Frommel, C., Leser, U.: Columba: an integrated database of proteins, structures, and annotations. BMC Bioinformatics 6 (2005)

13. Calvanese, D., Giacomo, G.D., Lenzerini, M., Naradi, D., Rosati, R.: Source integration in data warehousing. In: DEXA Workshop, pp. 192–197 (1998)

14. Kanehisa, M., Goto, S., Hattori, M., Aoki-Kinoshita, K.F., Itoh, M., Kawashima, S., Katayama, T., Araki, M., Hirakawa, M.: From genomics to chemical genomics: new developments in KEGG. Nucleic Acids Research 34, 354–357 (2006)

15. Chang, A., Scheer, M., Grote, A., Schomburg, I., Schomburg, D.: BRENDA, AMENDA and FRENDA the enzyme information system: new content and tools in 2009. Nucleic Acids Research 37, 588–592 (2009)

16. The UniProt Consortium: The Universal Protein Resource (UniProt) in 2010. Nucleic Acids Research 38, 142–148 (2010)

17. Benson, D.A., Karsch-Mizrachi, L., Lipman, D.J., Ostell, J., Sayers, E.W.: GenBank. Nucleic Acids Research 38, 46–51 (2010)

18. Munch, R., Hiller, K., Barg, H., Heldt, D., Linz, S., Wingender, E., Jahn, D.: PRODORIC: prokaryotic database of gene regulation. Nucleic Acids Research 31, 266–269 (2003)

19. Navas-Delgado, I., Aldana-Montes, J.F.: Extending SD-Core for Ontology-based Data Integration. Journal of Universal Computer Science 15, 3201–3230 (2009)

20. Navas-Delgado, I., Roldan Garcia, M.M., Mazorra, D.D., Aldana-Montes, J.F.: Developing Data Services. In: The 17th Conference on Advanced Information System Engineering. Data Integration and the Semantic Web, DISWeb 2005, vol. 4, pp. 287–301 (2005)

21. Chniber, O., Kerzazi, A., Navas-Delgado, I., Aldana-Montes, J.F.: KOMF: The Khoas Ontology-based Mediator Framework NETTAB 2008: Bioinformatics Methods for Biomedical Complex System Applications (2008)

22. Navas-Delgado, I., Kerzazi, A., Chniber, O., Aldana-Montes, J.F.: A Semantic Middleware Applied to Molecular Biology. In: Meersman, R., Tari, Z., Herrero, P. (eds.) OTM-WS 2008. LNCS, vol. 5333, pp. 976–985. Springer, Heidelberg (2008)

23. Navas-Delgado, I., Aldana-Montes, J.F.: SD-Core: Generic Semantic Middleware Components for the Semantic Web. In: Lovrek, I., Howlett, R.J., Jain, L.C. (eds.) KES 2008, Part II. LNCS (LNAI), vol. 5178, pp. 617–622. Springer, Heidelberg (2008)

24. Hillebrand, G.G., Kanellakis, P.C., Mairson, H.G., Vardi, M.Y.: Undecidable Boundedness Problems for Datalog Programs. J. of Logic Programming 25, 163–190 (1995)

25. Jorg, T., Dessloch, S.: Towards generating ETL processes for increlental loading. In: Proc. Intl. Symposium on Database Engineering and Applications, vol. 299, pp. 101–110 (2008)

26. Neerincx, P.B.T., Leunissen, J.: Evolution of web services in bioinformatics. Brief. Bioinform. 6, 178–188 (2005)

27. Marrakchi, K., Briache, A., Kerzazi, A., Navas-Delgado, I., Aldana-Montes, J.F., Ettayebi, M., Lairini, K., Rossi Hassani, B.D.: PseudomonasDW Ontology: an ontology for Pseudomonas species. Technical report (ITI.10-2), Department of Computer Languages and Computing Science, Higher Technical School of Computer Science Engineering University of Málaga (2010)

The Cinderella of Biological Data Integration: Addressing Some of the Challenges of Entity and Relationship Mining from Patent Sources

Ithipol Suriyawongkul[1,2], Christopher Southan[2], and Sorel Muresan[2]

[1] Chalmers University of Technology, Gothenburg, Sweden
[2] Computational Chemistry Section, Global Compound Sciences, DECS,
AstraZeneca R&D, Mölndal, Sweden
ithsur@student.chalmers.se

Abstract. Most of the global corpus of medicinal chemistry data is only published in patents. However, extracting this from patent documents and subsequent integration with literature and database sources poses unique challenges. This work presents the investigation of an extensive full-text patent resource, including automated name-to-chemical structure conversion, licensed by Astra-Zeneca via a consortium arrangement with IBM. Our initial focus was identifying protein targets in patent titles linked to extracted bioactive compounds. We benchmarked target recognition strategies against target-assay-compound relationships manually curated from patents by GVKBIO. By analysis of word frequencies and protein names we assessed the false-negative problem of targets not specified in titles and false-positives from non-target proteins in titles. We also examined the time-signals for selected target and non-target names by year of patent publication. Our results exemplify problems and some solutions for extracting data from this source.

Keywords: Biomedical text mining, Patent information.

1 Introduction

Drug discovery research has collectively generated a large experimental data corpus that intersects with enzymology, structural biology, metabolic biochemistry and chemical biology. Over the last 20 years, driven mainly by high-throughput screening (HTS) and new targets from the human genome, this has resulted in a substantial increase in patent filings, from commercial and academic institutions, claiming novel compounds as potential new medicines. The value of this output is derived in part from the fact that, in a community sense, the inventors largely overlap with authors in scientific journal papers [1]. While there is a tradition of "patent-then-publish" in the pharmaceutical industry a recent survey suggested that only 6% of chemical structures in patents were appearing in medicinal chemistry journals [2].

Thus, while the extent of "data-in-common" between patents and journal papers is largely unknown, applications focused on medicinal chemistry constitute a valuable biological data source. An impression of size is given in the World Intellectual

P. Lambrix and G. Kemp (Eds.): DILS 2010, LNBI 6254, pp. 106–121, 2010.

Property Organization (WIPO) annual statistics of global pharmaceutical patents of about 12,000 unique (i.e. not including patent families) applications per year. These documents are publically available and their basic information be searched from the major patent office portals. There are also newer resources such as Patent Abstracts [3], CiteXplore [4] (both from the EBI), Free Patents Online [5], Google Patents [6] and other initiatives that extend the indexing to sections of full text and, in the case of SureChem [7] also utilize automated name > chemical structure conversion.

Despite this increasing accessibility, the scientific information in these patents still remains under-utilized by the academic community, thus justifying its description as a "Cinderella" data source. In contrast, the commercial sector has paid heavily to support a patent information brokerage industry that produces range of database products designed to meet their searching needs. However, this has hitherto been largely confined to manual extraction of the highest value sections of patent data.

1.1 Patents as a Data Source

A general description of patent applications in the context of the pharmaceutical and biotechnological industries is provided by Webber [8]. The basic sections are shown in Table 1.

Table 1. Anatomy of a patent

Subsection of the patent	Function
Title, date, inventors, affiliations and abstract	Indexed by the patent offices to facilitate searching
Background description	Context for understanding the invention
Statements of invention	Basis for claims and restrictions
Examples and Figures	Reduction to practice in support of the claims
Claims	Explicitly defines the scope of the invention

Within a major pharmaceutical company such as AstraZeneca (AZ) the patent landscape is monitored across many areas relevant to R&D. However, the sub-set that this article is focused on exemplify and claim novel chemical structures supported by activity modulation data against defined molecular targets *in vitro*. Efforts to identify these as soon as possible right after publication include accessing commercial databases. These target-centric patents are scrutinized for intersections with internal projects, disease area interests or marketed products. They are thus a key information source in the pharmaceutical industry not only for competitive intelligence but also for structure-activity relationship (SAR) data integration [9]. As a document type they have the advantage of describing inventions that constitute, by definition, new knowledge for drug discovery [10]. In addition, data may be disclosed in patents long before they are published in journals [11]. Nevertheless, compared to journal papers, patent documents have significant disadvantages. These include 1) the competitive incentive of applicants to minimize or even obfuscate information, 2) lack of discussion on how the invention works, 3) non-standard terminology, 4) using 'prophetic examples' never actually carried out [10,11], 5) embedding results within 100's of pages of irrelevant text, 6) reporting only qualitative rather than quantitative data, 7) a "shotgun" approach to claiming many potential diseases for treatment, 8) a complex but not universally standardized document structure 9) the use of Markush-type specifications for large classes of compounds and 10) data redundancy for applications grouped within the same patent family. These challenges are exemplified by the observation that drug discovery project teams may spend several man-days combing out the details from a large patent, much longer that required for a typical medicinal chemistry journal paper.

1.2 Purposes of the Study

The aim of this work was to explore possibilities for automating retrieval of patents of the type described above. The initial focus was on the problem of identifying targets in patent titles that could be linked to bioactive chemistry and included benchmarking automated recognition strategies against manually curated patents. While it had already been observed that protein names in titles are likely to represent targets no systematic study of this has been reported.

2 Methods

2.1 Data Sources

2.1.1 The IBM Full-Text Patent Resources

Before the initiation of this study AstraZeneca entered into a consortium agreement with the IBM Almaden Research Center to access the IBM patent full-text extraction resources. This included on-line access to their Strategic Information Mining PLatform for IP Excellence (SIMPLE) [12]. This consists of a set of analytical tools that operate on a data warehouse containing full-text documents extracted and transformed via XML feeds from three patent offices including USPTO, EPO and WIPO. These are indexed by patent number, application date, publication date, inventors, assignee and International Patent Classification codes. The complete text is divided into title, abstract, claims and description. Both trivial and IUPAC systematic chemical names are converted into SMILES codes using the name=struct® program from CambridgeSoft and different synonyms mapped to the same structure [13].

SIMPLE includes entity annotators for chemical names, gene names and other biological terms that can be analyzed in combination with text searching [12,14]. The results presented were generated via queries of an Oracle instance of the IBM full text and extracted chemical structures database inside AstraZeneca but, during the period of this study, did not yet include non-chemical entity annotations. This AZ instance contained ~11.1 million documents, ~2.2 million with claimed chemical structures. Of these ~1.5 million SMILES codes in their claim sections were unique. This data source will henceforth be referred to as "IBM".

2.1.2 The GVKBIO Patent Database

AZ also licenses a Target Inhibitor Database from GVKBIO that includes compound-to-assay-to-target relationships manually extracted by expert curators [15]. The data are organized around five key entities. A typical example would be a document "D" describing a biochemical assay "A" with quantitative result "R" establishing compound "C" as an inhibitor of protein "P" (i.e. a D-A-R-C-P relationship). In this study, a subset of the GVKBIO Target Inhibitor Database with 43,085 patent numbers was used that had exact document matches in IBM. This data source will henceforth be referred to simply as "GVKBIO".

2.2 Target Terminology

While the term "drug target" is well understood it can be ambiguous in the context of patent documents. We will therefore use the term *"bona fide* target" in this work to mean the protein whose reported activity modulation *in vitro* is proposed to be the therapeutic mechanism for novel chemicals claimed in the patent. For an example, (used later in this report) a patent may report IC50 data on novel inhibitors of renin and claim these for the treatment of hypertension (i.e. a D-A-R-C-P relationship where "R" is an IC50 and "P" is renin). The problem faced during both manual and automated data extraction is that there may be a number of different A-R-C-P relationships in the same document. For example, it is not uncommon for patents claiming renin inhibitors to report their cross-reactivity against other biochemically significant human aspartyl proteases such as Cathepsin D. *Vice versa* patents claiming Cathepsin D inhibitors for Alzheimer's disease or cancer may include cross-screening data against renin. To pragmatically differentiate these we will use the term "target" to designate any protein with chemical modulation data (i.e. an A-R-C-P relationship) and "non-target" to refer to a protein specified in the patent without chemical modulation data.

2.3 Database Queries

Both the IBM and GVKBIO data sources are Oracle 10g databases accessible by Structured Query Language (SQL). IBM contains full-text description (i.e. patent number, title, abstract, claim section and description) of all patents published in USPTO, EPO and WIPO. GVKBIO contains target protein names manually annotated

Table 2. Experimental Objectives and Sources

Result Section	Description	Assessment	Search corpus	Reference corpus	Protein names
3.1	Manual identification of target protein names in titles	Recall	GVKBIO	-	50 target proteins
3.2	Automatic identification of target protein names in titles	Recall & Precision	IBM	GVKBIO	2 target proteins 2 non-target proteins
3.3	Using of chemical modulation keywords in titles to select patents with target proteins	Recall & Precision	IBM	GVKBIO	2 target proteins 2 non-target proteins
3.4	Assessment of improved recall from extending search to abstracts and claim sections	Recall	GVKBIO	-	2 target proteins
3.5	Assessment of potential false positives from extending search to abstracts and claim sections	Recall & Precision	IBM	GVKBIO	2 target proteins 2 non-target proteins

tagged for selected pharmaceutical patents resulting in patent-number-to-target relationships. Therefore, data for each patent document in these two sources can be linked by the patent number. For results shown in this article, all full-text patents were accessed from IBM. Selected results were calibrated via the target protein names curated by GVKBIO for the same patents, as indicated. Our investigations are categorized in Table 2 with detailed explanations given in each result section.

3   Results and Discussion

3.1   Manual Identification of Target Protein Names in Titles

To retrieve patents with specific target proteins it was necessary to manually assessment of the frequency of *bona fide* target names in titles. For this a set of 211 titles were selected from GVKBIO with the following criteria 1) they were published between years 2008 to 2010, 2) contained only one annotated human protein 3) the patent numbers were indexed in at least one other patent data source that AZ licenses. Of these, 171 unique titles encompassed 50 different target proteins (as curated by GVKBIO). There were many cases of identical titles for different patent numbers with the same applicants but not in the same patent families (e.g. WO2008116107 and WO200807978, both are GSK patents with title "glucokinase activators"). Each of the 171 titles was read manually by one of us with sufficient domain knowledge to recognize *bona fide* targets. A categorization of these is shown in Table 3 and the results of the assessment in Table 4.

Table 3. Classification of target protein recognition in titles

Categories	Descriptions
Positive	*Bona fide* target protein names (e.g. WO2008000409 "new CXCR2 inhibitors" or WO2009076337 "gamma secretase modulator")
Ambiguous	Generic or aggregate target designations that cannot be linked to a protein sequence (e.g. WO2008081208 "piperidine GPCR agonists" or WO2008115369 "derivatives of 5-amino-4,6-disubstituted indole and WO2009023677 "5-amino-4,6-disubstituted indoline as potassium channel modulators")
Negative	No target protein name (e.g. WO2009059961 "a method of hormone suppression in humans" or WO2009010794 "2,4-diamino-pyrimindine derivatives")

Table 4. Assessment of expert recognition of target protein names in titles

Categories	Number of unique patent titles	
Positive	87	(51%)
Ambiguous	15	(9%)
Negative	69	(40%)
Total	171	(100.0%)

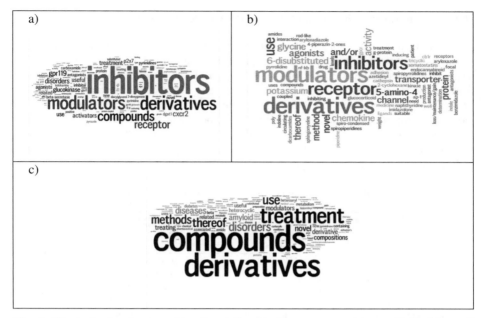

Fig. 1. Word clouds of patent titles for a) "positive" target recognition , b) "ambiguous" and c) "negative"

We extended the analysis by using Wordle [16] to visually display the term frequencies in the three categories of titles from Table 3 (Figure 1).

We can see (Figures 1 a) and b) that the "positive" and "ambiguous" categories contain keywords associated with chemical bioactivity (e.g. "inhibitor" and "modulator"). Retrieval of these patents by combining these keywords can be achieved with a recall approaching 60%. Note some *bona fide* targets are producing a signal in the title word clouds (e.g. GPR 119, P2X7 and CXCR2).

We can also detect (Figure 1 c) that the "negative" category usually includes associated disease and/or chemical series names (e.g. "treatment", "disease", "compounds" and "derivatives"). This category also includes cases where recognizable protein names in titles are not *bona fide* targets (e.g. US20080280948 with title "modulator of amyloid beta" claims gamma secretase inhibitors rather than amyloid disaggregation). The association is thus interpretable but the target name is, strictly speaking, a false positive. It is important to emphasize that this experiment was performed on a set of sample patents from GVKBIO which had a target protein identified from each patent via manual curation and cannot therefore predict the precision for retrieving patents using occurrence of protein names in titles. It is also not possible to extrapolate to equivalent precision using a corpus of patents where the presence of a *bona fide* target is unknown.

3.2 Automatic Identification of Target Protein Names in Titles

The results in the previous section suggested that recall for automated retrieval of *bona-fide*-target-containing patents via protein names in titles would be acceptable, especially considering the size of the IBM data. To avoid the problem of synonym variability, four protein names with relatively clean synonyms were chosen (Table 5). Renin and thrombin are *bona fide* target proteins, while albumin and hemoglobin are non-target controls. The data sets used in this experiment were IBM from 1980 - 2009 (10,846,899 patents) and a subset of GVKBIO from 1980 - 2009 (34,441 patents). Synonyms of the four names (Table 5) were searched against patent titles in IBM and calibrated against the same (target annotated) GVKBIO patents (Table 6).

Table 5. Protein names, HGNC symbols and their synonyms

Protein Names	HGNC Symbols	Synonyms
Renin	REN	rennin angiotensinogenase
Thrombin	F2	prothrombin coagulation factor II coagulation factor 2
Albumin	ALB	serum albumin
Hemoglobin	(many subunits)	haemoglobin

Table 6. Retrieved patents using protein synonyms in titles

Protein Names	Number of Patents Retrieved			Approx. % Recall	Min. % Precision
	GVKBIO	IBM	patents-in-common betw. GVKBIO & IBM		
Renin	494	813	237	48.0%	29.2%
Thrombin	890	1743	215	24.2%	12.3%
Albumin	5	1200	0	0.0%	0.0%
Hemoglobin	0	1542	0	-	0.0%

For the results in Table 6, it should be noted that GVKBIO does not cover all patents for a particular target protein. Therefore, patents-in-common between GVKBIO and IBM cannot be used to infer the number of true positives obtained directly. Nevertheless, this can be interpreted as the minimum number of true positive which can then be used to estimate the minimum precision as shown (Formula 1, below). For renin, there are 813 patents in the IBM search result with 237 in-common with GVKBIO (with renin as a target protein). Therefore, the minimum precision of this approach can be estimated as 29%

$$\% \ Precision = \frac{TP}{TP+FP} \geq \frac{Number \ of \ common \ patents \ between \ GVKBIO \ and \ Search \ Result}{Number \ of \ patents \ in \ the \ Search \ Result} \quad (1)$$

$$\% \ Recall = \frac{TP}{TP+FN} \approx \frac{Number \ of \ common \ patents \ between \ GVKBIO \ and \ Search \ Result}{Number \ of \ GVKBIO \ curated \ patents \ for \ a \ particular \ protein} \quad (2)$$

By the same reasoning, the number of false negatives cannot be obtained directly by calibrating the search result against the GVKBIO, but the minimum number of false negatives can. However, since GVKBIO represents a sample set of all patents with target proteins, an approximate percentage recall can be obtained by using Formula 2 (above). In the case of renin, there are 494 patents in GVKBIO, and 237 of them have renin in titles. Therefore, approximate percentage recall of this retrieval approach would be 48.0%.

Results show different precision between the two targets. Some of this is due to search specificity noise (e.g. kynurenine or antithrombin). This type of false-positive could potentially be removed by using more advanced term recognition rules. For thrombin, some false negatives (w.r.t. GVKBIO) are where some *bona fide* targets for those specific patent documents are, for example, factor Xa or factor VII. (e.g. US20080214495 "heterocyclic sulfonamide derivatives as inhibitors of factor Xa" and US7576098 "heterocyclic compounds as inhibitors of factor VIIa"). The inclusion of cross-screening data against thrombin is not unexpected. This exemplifies cases in which there are data for multiple target proteins in a patent (e.g. factor Xa and thrombin), but the *bona fide* target of the patent may be only one (i.e. factor Xa). We also detected examples of combined targets (e.g. WO2004052851A1, "pyrrolydin-2-one derivatives as inhibitors of thrombin and factor Xa" and EP1294684A2 "thrombin or factor Xa inhibitors") but it is not clear if these represent authentic polypharmacolgy or claim "bet hedging".

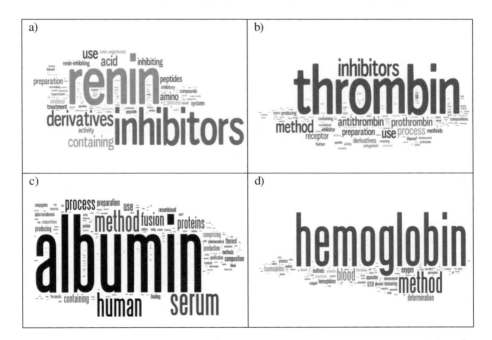

Fig. 2. Word clouds of patent titles obtained by searching synonyms of non-target proteins in titles. a) renin, b) thrombin, c) albumin, d) hemoglobin.

Manually skimming the 1200 patent titles with albumin and 1542 for hemoglobin indicated they were probably all, in the *bona fide* target sense, false positives. For albumin many claim methods of conjugation with other proteins (e.g. WO2009121884A1 "insulin albumin conjugates") . Note that Table 6 includes five GVKBIO patents with albumin as a target name. While they could be classified as false-positives in fact they are cases where the chemical modulators of the *bon fide* targets have also been tested for albumin binding (e.g. EP1586318 "thiadiazolidi-nones as GSK-3 inhibitors"). This exemplifies another challenge for recognition and extraction of target proteins from patents. Most of the hemoglobin applications spec-ify analytical methods (e.g. EP2016390A1 "a method and a system for quantitative hemoglobin determination"). Supporting the non-target inferences are the very low frequencies of chemical modulation keywords (e.g. "inhibitor", "modulator", etc.) in titles. Word clouds of the titles retrieved with the four proteins are shown in Figure 2.

One of the conclusions we can draw from these results is that using protein names and synonyms alone to search against patent titles is likely to have a high false posi-tive rate for *bona fide* targets. However, we explored another approach as a partial solution. Our experience with mining GVKBIO data was that *bona fide* targets often show a "time signal" in the sense that the generation rate of published data directly associated with these targets can vary significantly on a year-to-year basis. There are many possible causes of these fluctuations that are difficult to verify formally. How-ever, it is known that declared target success milestones (e.g. new validation data,

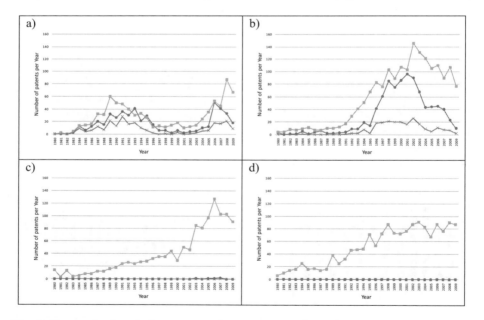

Fig. 3. The result of retrieving patents using occurrence of protein synonyms in titles for a) Renin, b) Thrombin, c) Albumin, d) Hemoglobin. (━●━ represents number of GVKBIO pat-ents tagged with a particular protein name, ━■━ represents number of IBM patents obtained from search, ━✕━ represents number of patents matched between GVKBIO and IBM search results)

initiation of clinical trials or an NCE submission) invariably trigger some level of follow-on activity that can result in a subsequent "spike" of new patent applications. To test this we have plotted the frequency of the four proteins in patent titles from 1980 to 2008 (Figure 3).

It can be seen (Figure 3 c) and d)) that the two non-targets show a steady increase. Nonetheless there is suggestion of a peak for albumin at 2006 -2007, although as already explained this may be new-use related but not target related. In contrast, the two targets (Figure 3 a) and b)) not only show strong signals but that these appear to be correlated between GVKBIO and IBM. As these are selected by curated target in the former case this suggests the signal in the latter case may be authentic in representing a significant increase in patent publications for these *bone fide* targets.

3.3   Using of Chemical Modulation Keywords in Titles to Select Patents with Target Proteins

It has been shown in the section 3.2 that retrieving of patents with target proteins simply by searching occurrence of protein synonyms in titles could result in many false positives. However, patent titles within these false positives also have a very low occurrence of keywords signifying chemical modulation (Figure 2). In contrast, patent titles within true positives are having high co-occurrence of target protein names and these keywords in titles (Figure 1.a). This led to an initiative to retrieve patents with *bona fide* target proteins by searching for the co-occurrence of protein names with these keywords in titles, which could possibly produces better precision (i.e. false positives removed) while sustaining percentage recall. A set of keywords were selected from a corpus of titles of patents with target proteins. These keywords were then applied to each search result set obtained in section 3.2 to filter out patent titles without these keywords.

3.3.1   Evaluation of Chemical Modulation Keywords

In order to get a set of keywords signifying probable chemical modulation of targets, a word frequency analysis was performed. This used the 34,575 GVKBIO patents used in section 3.2 but expanded to included patents published between years 1973-2009. The result shows these titles include 16,714 word forms, excluding stop words [17]. Those signifying chemical modulation were classified into four groups (Table 7.a) and their frequencies infer the approximate numbers of patent titles matched by these keywords.

To estimate recall the search terms (Table 7.b) were derived by stemming their plural forms. Each keyword group was then assessed by searching its derived search terms against the set of 34,575 patent titles (Table 7.b). Note that the number of patent titles matched with each keyword group could be different from the corresponding approximate number in shown in table 7.a. This is because keywords in each group match multiple keyword variants (e.g. "agonist-specific", "tumor-inhibiting" and "immunomodulators"). Note that the "agonism" keyword group gave a higher than expected recall due to substring matching between the term "agonist" and occurrence of "antagonist" in titles. This is an inherent problem when trying to differentiate nested terms. The results show these keywords can be used collectively to retrieve

Table 7. Chemical modulation keywords extracted from patents with target proteins (GVKBIO patent database). a) word frequency analysis, b) retrieval testing.

Keyword Groups	(a) Word frequency analysis		(b) Retrieval testing using keywords	
	Keywords as found in patent titles	Occurrence frequency (out of 34,575 titles)	Derived search terms used in retrieval	Retrieval result (# of patents)
Agonism	agonism, agonist(s), agonistic, agonist-like	1775 (5.1%)	agonism, agonist, agonistic	6270 (18.1%)
Antagonisation	antagonisation, antagonise, antagonising, antagonism, antagonist(s), antagonistic(s), antagonize, antagonizing, antagonist-like	4702 (13.6%)	antagonisation, antagonise, antagonising, antagonism, antagonist, antagonistic, antagonize, antagonizing	4782 (13.8%)
Inhibition	inhibit(s), inhibiting, inhibition, inhibitive, inhibitor(s), inhibitory	8756 (25.3%)	inhibit, inhibiting, inhibition, inhibitive, inhibitor, inhibitory	8762 (25.3%)
Modulation	modulate(s), modulating, modulation(s), modulator(s), modulatory	2624 (7.6%)	modulate, modulating, modulation, modulator, modulatory	2640 (7.6%)
	Total (Summation)	17857 (51.6 %)	Union (all keywords)	17503 (50.6%)

over 50% of patents with protein chemical modulation, regardless of protein name occurrence i titles. Extending the list to include more terms (e.g. "activation") might further improve recall.

3.3.2 Testing Chemical Modulation Keywords

We tested co-occurrence of a protein name and chemical modulation keywords in titles to filter-out patents without *bona fide* targets. The set of keywords obtained in section 3.3.1 were applied to each result obtained in section 3.2 In effect, the search result after filtration contains only patents with co-occurrence of the protein name and keywords in titles as shown in Figure 4. The resulting recall and precision were compared to using only the protein name from section 3.2 as shown in Table 8.

The result shows that the chemical modulation keywords in titles can be used to filter out patents with non-target protein names in titles and thereby improve precision for both target and non-target proteins. For albumin this keyword filtration removed nearly 1200 false positives (Table 8, Figure 3.c). An advantage of this approach is that it does not appear to significantly degrade recall. For example, 237 patents are recalled by using co-occurrence of renin synonyms and chemical modulation keywords, which is the same number of patents as recalled by using only the renin synonyms

Table 8. Patents retrieved from IBM using a) only protein names and synonyms, b) combining these with chemical modulation keywords

Protein Names	GVKBIO	(a) Without filtration by keywords				(b) With filtration by keywords			
		Number of Patents		Approx % Recall	Min % Precision	Number of Patents		Approx % Recall	Min % Precision
		IBM Search Result	Match btw. GVKBIO & IBM Search Result			IBM Search Result	Match btw. GVKBIO & IBM Search Result		
Renin	494	813	237	48.0%	29.2%	723	237	48.0%	32.8%
Thrombin	890	1743	215	24.2%	12.3%	903	208	23.4%	23.0%
Albumin	5	1200	0	0.0%	0.0%	21	0	0.0%	0.0%
Hemoglobin	0	1542	0	-	0.0%	15	0	-	0.0%

(Table 8, Figure 3.a). This implies that patent titles with target protein names often include these keywords.

The utility of this combination (Figure 4) is also demonstrated when applied to the longitudinal analysis already shown in Figure 3. The use of the combination is thus very effective at filtering out the non-target proteins but maintaining the signals of *bona fide* target names.

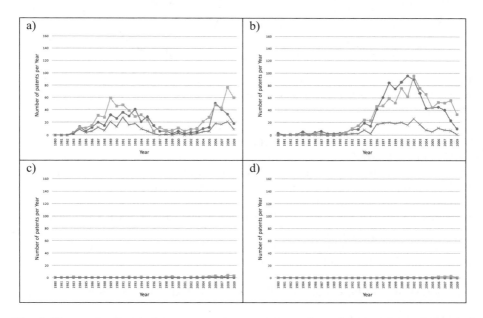

Fig. 4. The result of retrieving patents using occurrence of protein synonyms and chemical modulation keywords in titles for a) Renin, b) Thrombin, c) Albumin, d) Hemoglobin. (—●— number of GVKBIO curated patents, —■— number of IBM search result, —✕— number of match between GVKBIO and IBM)

3.4   Assessment of Improved Recall by Extending Search to Abstracts and Claims

In this section we tested extending the searching for protein synonyms in other sections of the text to retrieve patents without target names in titles. From the same set of GVKBIO used in section 3.2 (34,411 patents), we selected 8,167 patents published between years 2006 – 2009 where IBM had the full-text documents (these more recent dates showed a better quality of text extraction for abstracts and claims due to direct XML feeds from the patent offices). Within two sets of 79 renin and 80 thrombin patents, we counted synonyms in titles, abstracts and claim sections, respectively (Figure 5).

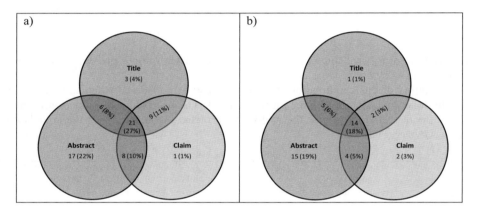

Fig. 5. Venn diagram for retrieved results from a set of patents containing a particular target protein (retrieved by searching for the protein name in titles, abstracts and claim sections) for a) Renin, b) Thrombin

For renin patents (Figure 5.a), searching for the protein synonym could retrieve 39 (49%) in title, 52 (66%) in abstract, 39 (49%) in claim section, and 65 patents (82%) in all three sections. Interestingly, searching in abstracts retrieved 17 unique patents (22%) not found by searching in titles or claim sections. The equivalent figures for the thrombin patents (Figure 5.b) were 22 (28%) in title, 38 (48%) in abstract, 22 (28%) in claims, and 43 patents (54%) in all sections. In this case abstracts were also shown to retrieve 15 unique patents (19%). Thus, extending searched from titles to abstracts and claim sections could significantly improve patent target retrieval coverage, mainly via unique information in abstracts.

3.5   Assessment of Potential False Positives from Extending Section Searching

To examine false-positives we used the same target and non target pairs to search titles, abstracts and claim sections for IBM 2006 – 2009 USPTO patents (1,234,684). Results for each protein name were then calibrated against the same USPTO patents found in the 34,441 GVKBIO set used in section 3.2. (Figure 6).

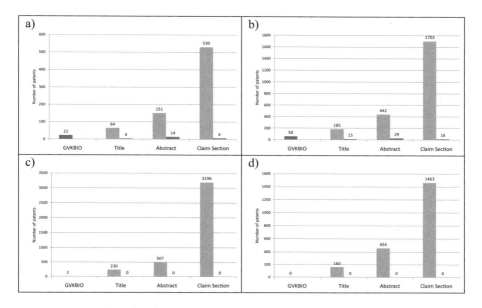

Fig. 6. The result of retrieving patents with a particular target protein using occurrence of the protein names in titles, abstracts and claim sections for a) renin, b) thrombin, c) albumin, d) hemoglobin. (━●━ number of GVKBIO curated patents, ━■━ number of IBM search result, ━✕━ number of match between GVKBIO and IBM)

Results from albumin and hemoglobin (Figure 6.c and d) suggest that searching protein names in abstracts and claims leads to substantial increases in false positives in claims. Similarly, thombin and renin (Figure 6.a, Figure 6.b) also show substantial increase of patents unmatched by GVKBIO, which could be potential false-positives. Comparing the matches between IBM and GVKBIO suggests abstracts show the highest retrieval in line with the results from section 3.4.

4   Conclusions

This work illustrates some of the challenges and possible solutions for mining drug discovery data from full-text patents. The results are preliminary in that we have initially focused on the problem of recognizing target names. We have shown that protein names in titles can be used to retrieve patents with probable *bona fide* targets and claimed bioactive chemical structures. We also demonstrated that recall is protein-specific and results for non-target proteins can generate false positives. However, combining protein names and chemical modulation keywords for searching titles improved precision without significant loss of recall. Extending this type of search from titles to other sections significantly improved target retrieval via the extra information in abstracts with the caveat of increased false positives from claims. Therefore, combining title with abstract searches represented a useful compromise.

As has often been shown in the field of bioinformatics the value of expert manual data extraction proves to be complementary to automated extraction rather than competitive. In addition, the former provides definitive test corpora for the latter, as we have demonstrated using GVKBIO curation to benchmark IBM. The ability of a human expert to scan a patent, identify the *bona fide* target, assay descriptions, results, drug-like structures, data relationships and mechanistic concepts cannot yet be matched by automated text mining. However, access to a substantial corpus of full-text patents from which over a million of unique claimed chemical structures have been extracted by the IBM process is clearly of complementary value for its scale, coverage and speed that substantially exceeds the capacity of manual efforts.

We can envisage several ways forward to converge the two approaches. There are a number of recent advances in text mining for recognizing and extracting chemical and biological information from non-patent literature, some of which could be directly applied to extending this work [18,19,20]. Natural Language Processing (NLP) seems a good candidate because of its ability to discriminate context (e.g. for targets and inhibitors) beyond just the recognition of protein names. However, such new approaches will need to be tested and adapted (e.g. via the optimization of rule sets) because of the major differences between papers and patents already described. One of the utilities envisaged from the approaches we have begun to explore here would be automated alerts for new D-A-R-C-P relationships, thus facilitating integration of patent data with other sources to extend not only the drug target landscape but also the concomitant bioactive chemical space.

References

1. Agarwal, P., Searls, D.: Can literature analysis identify innovation drivers in drug discovery? Nature reviews. Drug discovery 8, 865–878 (2009)
2. Southan, C., Várkonyi, P., Muresan, S.: Quantitative assessment of the expanding complementarity between public and commercial databases of bioactive compounds. Journal of cheminformatics 1, 10 (2009)
3. Patent Abstracts, http://srs.ebi.ac.uk
4. CiteXplore, http://www.ebi.ac.uk/citexplore
5. Free Patents Online, http://www.freepatentsonline.com
6. Google Patents, http://www.google.com/patents
7. SureChem, http://www.surechem.org
8. Webber, P.: A guide to drug discovery. Protecting your inventions: the patent system. Nature reviews. Drug discovery 2, 823–830 (2003)
9. Grandjean, N., Charpiot, B., Pena, C., Peitsch, M.: Competitive intelligence and patent analysis in drug discovery: Mining the competitive knowledge bases and patents. Drug Discovery Today: Technologies 2, 211–215 (2005)
10. Granstrand, O.: The economics and managment of intellectual property: towards intellectual capitalism. Edward Elgar Publishing Limited (2000)
11. Grubb, P.: Patents for chemicals, phamaceuticals, and biotechnology. Oxford Univ. Press, New York (2004)
12. Chen, Y., Spangler, S., Kreulen, J., Boyer, S., Griffin, T.D., Alba, A., Behal, A., He, B., Kato, L., Lelescu, A., Zhang, L., Kieliszewski, C.: SIMPLE: A Strategic Information Mining Platform for IP Excellence, San Jose, CA, USA (2009)

13. Brecher, J.: Name=struct: A practical approach to the sorry state of real-life chemical no-menclature. Journal of Chemical Information and Computer Science 39, 943–950 (1999)
14. Rhodes, J., Boyer, S., Kreulen, J., Chen, Y., Ordonez, P.: Mining patents using molecular similarity search. In: Pacific Symposium on Biocomputing 2007, Maui, Hawaii, p. 304 (2007)
15. Sarma, J., Radha, K.: Database systems for knowledge-based discovery. In: Chemogenom-ics: Methods and Applications, vol. 575, pp. 159–172 (2009)
16. Wordle, http://www.wordle.net
17. Stop words. Department of Computer Science, Cornell Univesity, ftp://ftp.cs.cornell.edu/pub/smart/english.stop
18. Banville, D.: Mining chemical and biological information from the drug literature. Current Opinion in Drug Discovery & Development 12(3), 376–387 (2009)
19. Krallinger, M., Morgan, A., Smith, L., Leitner, F., Tanabe, L., Wilbur, J., Hirschman, L., Valencia, A.: Evaluation of text-mining systems for biology: overview of the Second Bio-Creative community challenge. Genome Biology 9, S1 (2008)
20. Cohen, A., Hersh, W.: A survey of current work in biomedical text mining. Briefings in Bioinformatics 6(2), 57–71 (2004)

Algorithm for Grounding Mutation Mentions from Text to Protein Sequences

Jonas Bergman Laurila[1,*], Rajaraman Kanagasabai[2,*],
and Christopher J.O. Baker[1,**]

[1] University of New Brunswick, Saint John, New Brunswick, Canada
j02h9@unb.ca, bakerc@unb.ca
[2] Institute for Infocomm Research, Singapore
kanagasa@i2r.a-star.edu.sg

Abstract. Protein mutations derived from in vitro experimental analysis are described in detail in scientific papers. Reuse of mutation impact annotations is an important subfield of bioinformatics for which mutation grounding is a critical step. Presented here is a method for grounding of textual mentions from papers describing mutational changes to proteins. We distinguish between grounding of mutation entities to protein database identifiers and to the correct positions on sequences extracted from protein databases. The grounding workflow coordinates the extraction of mutation, protein and organism mentions from texts and uses these to identify target sequences. Mutation mentions are sequentially mapped onto candidate proteins to facilitate their correct grounding to a protein sequence, independent of a protein-mutation tuple extraction task. Using a gold standard corpus of full text articles and corresponding protein sequences we show high performance precision and recall and discuss novel aspects of the algorithm in the context of previous work.

Keywords: Natural Language Processing, Mutation Extraction, Mutation Grounding, Sequence Analysis.

1 Introduction

Mutations are variations in the genomic material of living organisms which can be caused by many different factors; viruses, radiation and mutagenic chemicals to mention a few. When these variations are nonsynonymous, point mutations will be introduced in the translated amino acid sequence, which can impact the protein properties, e.g. function or stability and ultimately be a cause for diseases. Researchers in molecular biology and medicine have created an enormous number of observations which are now distributed across the scientific literature or small gene- or organism-specific databases.

Attempts have been made to organize the scattered information. For example, the Human Genome Variation Society has the mandate to "*ensure documentation,*

---

* Both authors contributed equally to this work.
** Corresponding author.

P. Lambrix and G. Kemp (Eds.): DILS 2010, LNBI 6254, pp. 122–131, 2010.
© Springer-Verlag Berlin Heidelberg 2010

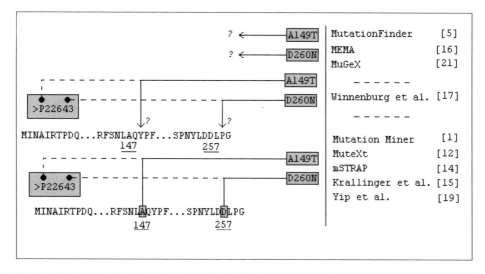

Fig. 1. Degrees of Mutation Grounding. *Uppermost,* mutation mentions are extracted but their relation to the appropriate protein sequence is not (no grounding). *Middle,* the related protein is found and mappings are established to the sequence if both text and database follow the same numbering scheme. *Lowermost,* the mutations are properly grounded, i.e. mapped to the correct position on the amino acid sequence of the related protein. Systems performing and not performing mutation grounding are displayed to the right.

collection, and free distribution of all variation information." [11]. In addition to having their own journal, Human Mutation, they keep an updated list of variation databases on their webpage[1].

Since the completion of the human genome sequencing project, and emergence of new methods for detecting sequence variations, the amount of information keep on growing at a rate that is too high for manual database insertion and adequate curation. To address this problem numerous systems for mutation extraction have been designed [17,15,14,5,19,1,12,16,21] with some [17,15,14,19,1,12] performing mutation grounding, which is the task of mapping extracted mutation mentions positions correct to sequences already stored in sequence databases (See Fig. 1). There are differing approaches to mutation grounding, each with their own shortcomings and interpretation of the task.

Mutation grounding is a prerequisite for many types of biological data integration, including visualization of annotated protein structures [9,13,1,14], pathway analysis, simulation and modelling purposes [2] or a required step in the prediction of stabilizing regions in membrane proteins [17]. Following mutation grounding, some of these approaches also depend on NLP tasks for extraction of impacts of mutations.

---

[1] http://www.hgvs.org/

1.1   Content Overview

The paper is outlined as follows. Section 2 covers the initial methods for entity recognition. Section 3 handles our method for mutation grounding. Section 4 present evaluations performed on our test corpora. The last section discuss how well the algorithm performed the mutation grounding task, together with comparisons to related work.

2   Prerequisites for Mutation Grounding

Before grounding of mutation mentions is possible, it is necessary to find all such mentions throughout the text together with other named entities. Typically a mutation extraction system mainly deals with three different entity recognition tasks; protein, organism and mutation recognition. According to [18] these tasks are defined as (i), in the case of proteins and organisms; to identify each occurrence of protein- and organism names and abbreviations and (ii) in the case of mutations; to detect each description of a mutational change described in a document, including the abbreviated and full-text forms. Moreover, nominal and pronominal coreferencing should also be detected.

2.1   Method for Entity Recognition

For the entity recognition task we use GATE [7] in combination with custom curated gazetteer lists created as described below. In our approach we do not solve pronominal coreferencing however nominal coreferencing for proteins and mutations are solved by the subsequent grounding step.

Protein and Gene Names. The protein database Swiss-Prot, a manually annotated part of UniProt Knowledge Base [3], was used to select protein and gene names. The use of Swiss-Prot is motivated by their high quality namings and mappings between names and protein sequences. The text format version of Swiss-Prot was encoded into our Mutation Grounding Gazetteer lists (MGG) and Mutation Grounding Database (MGDB). Separate gazetteer lists were made for those with names containing only one word and those containing names with more than one word. This separation is made to increase precision of named entity recognition by using case sensitive matching for shorter names and to increase recall by using case insensitive matching for longer names. Mappings between names and primary accession numbers were exported, together with mappings between primary accession numbers and amino acid sequences to the MGDB.

Organism Names. Swiss-Prot is used to extract names, both scientific (Latin genus and species) and English names. As a result one single gazetteer list containing all organism names is created and mappings between synonyms in English and scientific names are exported together with mappings between scientific names and primary accession numbers to the MGDB.

Mutation Descriptions. To extract mutation mentions we used the Mutation-Finder system [5]. The system employs a complex set of regular expressions and

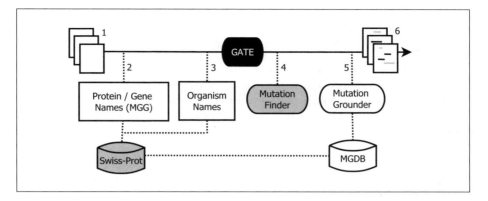

Fig. 2. Entity Recognition and Mutation Grounding Framework. Full-text documents (1) are run through a GATE pipeline with gazetteers derived from Swiss-Prot (2,3) and created with MutationFinder (4). Mutations and proteins can then be grounded (5). The output consists of annotated text (6).

is currently the best available tool for point mutation extraction. Full-text documents are first run through MutationFinder to create gazetteer lists containing mutation mentions that are compliant with the GATE framework. Mutation-Finder is also able to normalize mutations into wNm format, where w and m are one-letter codes for the wildtype and mutation residues, and N is the position on the amino acid sequence. E.g. normalize the mutation in "*In order to further understand the catalytic mechanism we constructed an Asp-124→Asn mutant enzyme*" to D124N. Normalization is required prior to the mutation grounding task.

Summary: full-text documents are run through a GATE pipeline containing our Mutation Grounding Gazetteer lists with organism-, protein/gene- and mutation terms[2]. The annotated mentions can now be used for further analysis in the protein and mutation grounding step. See figure 2 for a workflow overview.

3   Protein and Mutation Grounding

Protein grounding (GR-1) is defined as "*Assign[ing] the correct UniProtKB ID to each detected protein entity.*" and mutation grounding (GR-3) as "*Verify[ing] and, if necessary, positionally correct each mutation location to match its corresponding protein's sequence as obtained from UniProt KB*" [18].

3.1   Method for Mutation Grounding

The method described herein combines the protein grounding- and mutation grounding tasks into one single task, where information from protein mentions

---

[2] 20915 organism terms, 105012 protein/gene terms.

is used to enhance mutation grounding and vice versa. The method is based on the following assumptions:

A) Most papers concerning mutations are mainly about one protein or gene at a time[3].

B) The protein described in the document is likely to be mentioned most frequently, as a protein name, abbreviation or the corresponding gene name.

C) Organisms may or may not be mentioned throughout the text. Occurring organism mentions could be related to the expression host, rather than the source organism.

1: **function** GROUND(M, $sequence$)   ▷ M - set of mutations, $sequence$ - amino acid
   sequence
2:    $GSet \leftarrow \varnothing$
3:    **for all** $m_i \in M$ **do**
4:        **for all** $m_j \in M \setminus \{m_i\}$ **do**
5:            $regex \leftarrow$ BUILDREGEX(m_i, m_j)
6:            **while** MATCH($regex$, $sequence$) **do**
7:                $displacement \leftarrow$ CALCULATEDISPLACEMENT($regex$, $sequence$)
8:                $G \leftarrow \{m_i, m_j\}$
9:                **for all** $m_k \in M \setminus G$ **do**
10:                    **if** $sequence[m_k.position - displacement] = m_k.wildtype$ **then**
11:                        $G \leftarrow G \cup \{m_k\}$
12:                    **end if**
13:                **end for**
14:                $GSet \leftarrow GSet \cup \{G\}$
15:            **end while**
16:        **end for**
17:    **end for**
18:    $GSet \leftarrow$ MOSTMUTATIONS($GSet$)
19:    **return** LEASTDISPLACEMENT($GSet$)
20: **end function**

Fig. 3. Mutation Grounding algorithm. A set of point mutations are mapped together on an amino acid sequence, starting from minimum amount of mutations (two) and extending with as many mutations as possible one by one.

These assumptions give us an advantage, as sentence level co-occurrence or other prelusive methods of relation detection may be unnecessary. Our method works in the following manner.

1) For each document, retrieve all gene- and protein mentions.

2) For each gene- and protein mention m, retrieve all related accession numbers from the MGDB. An accession number is considered related if it was previously mapped to a name equal to m*, e.g. if "Carbonic Anhydrase" is found,

---

[3] In our test corpus, the amount of articles mainly about one single protein was just above 86%.

accession numbers related to Carbonic Anhydrase I, Carbonic Anhydrase II etc will be retrieved. This is because in some cases, authors tend to use shorter names when they have already stated the correct name. The retrieved accession numbers are pooled.

3) All pools are put together and all but the most frequently occurring accession numbers are discarded.

4) Retrieve all organism mentions and remove accession numbers that are not related, based on the mappings stored in the MGDB, to any of the retrieved organisms. If this will result in an empty pool of accession numbers, the changes from this step are discarded.

5) Retrieve all unique mutation mentions, normalize with MutationFinder and try to fit as many as possible onto the sequences corresponding to the accession numbers still left. The mutation grounding algorithm displayed in figure 3 makes use of regular expressions[4] similar to those used by [14]. These are used to handle possible differences in residue numbering, except that it only uses two mutations at a time as oppose to [14] who uses all mutations at the same time. This is to make it possible to ground mutations even if one or more of the set does not fit the sequence. When two mutation mentions fit the sequence, the remaining mutation mentions are tried one after another, taking in account the numbering displacement found when using the regular expression.

6) The accession number and corresponding sequence on to which most mutations are grounded is now considered as the correct one for the entire document. In case of a tie, the sequence with least displacement from the mutation numbering in the paper is chosen. Mutation mentions that do not fit the sequence are discarded. This will make sure that mutations of unrelated proteins mentioned in e.g. the reference section or type C proteins as [20] label them, in addition to mentions not describing point mutations at all.

7) The results are exported in form of annotations in the source documents, e.g. *" This patient demonstrated an* ⟨**Mutation ac=**"**O00255**" **correctPosition=**"**279**" **m =**"**A**" **w=**"**E**"⟩*E274A*⟨**/Mutation**⟩ *germ-line point mutation."*

4   Evaluation

To evaluate the method for mutation grounding a gold standard corpus was built using the COSMIC database [8]. Literature about haloalkane dehalogenases from the protein engineering domain was also used to show that the methods presented are not overfitted to papers describing disease causality.

In the COSMIC case three target proteins/genes were considered, PIK3CA, FGFR3 and MEN1. Full-text papers containing more than one single point mutation were chosen, with a total number of 73 documents. 63 of these documents were mainly about one single gene; the other 10 were therefore discarded. For

---

[4] For two normalized mutation mentions $w_1 N_1 m_1$ and $w_2 N_2 m_2$, sorted in the ascending order of N_i, the regular expression will be $w_1 \cdot \{N_2 - N_1\} w_2$. E.g. $A378C$ and $S381L$ will result in $A \cdots S$.

performance evaluation, we used two measures: precision which is defined as the fraction of correctly grounded mutations (true positives) over all grounded mutations (true positives + false positives) and recall which is defined as the fraction of correctly grounded mutations over all mentioned mutations (true positives + false negatives). A mutation is considered correctly grounded if it is mapped to the sequence related to the Swiss-Prot accession number displayed by COSMIC. A mutation is considered to be mentioned if it is found in COSMIC or if it is found by the MutationFinder system and validated by a domain expert. This last condition was created to avoid problems with mutations mentioned in text, referring to previous work, and therefore not found in COSMIC or type B proteins as labeled by [20]. The total number of unique[5] mutation mentions contained in the COSMIC corpora is 343.

The haloalkane corpus is an extension of the corpus used by [1] and contains 13 documents with a total of 54 unique mutation mentions. The documents were chosen on the same premises as in the COSMIC case, but as this corpus is not related to a database of mutations, all mutation mentions are found and validated by a domain expert.

The final corpus consist of 76 documents, were almost one third (25 documents) use a different numbering scheme than the sequences retrieved from Swiss-Prot and encoded in to the MGDB, i.e. in need of positional adjustment of the mentioned mutations during mutation grounding. Different numbering schemes can arise when authors trim signal peptides or other regions from the target sequences before carrying out site directed or other mutagenesis. Our evaluations are done one document at a time and the result is summed up per corpus and displayed in table 1.

Table 1. Mutation grounding performance

Corpus	Precision	Recall	Corpus size
PIK3CA	0.86	0.70	30
FGFR3	0.89	0.66	26
MEN1	0.54	0.32	7
Haloalkane Dehalogenases	0.83	0.73	13
Average[6]	0.84	0.65	76

5   Discussion

These results show that mutation grounding, with 84% precision, is mature enough to reliably be used to facillitate reuse of mutation annotations. The lower precision in the MEN1 case is due to usage of synonyms not included in

---

[5] Unique per document.

[6] We chose micro average, i.e. weighted by corpus size, since we have a significant difference in number of documents per corpus.

Swiss-Prot (MENI instead of MEN1) which results in the second most occurring protein to be chosen e.g. DNA-polymerase and ultimately incorrect grounding of mutations (false positives).

The overall recall is in line with typical performance of gazetteer based approaches but still not satisfying. In the PIK3CA and MEN1 cases the main reason for the low recall is because they occur as different isoforms, and the sequence retrieved by our system is always the main[7] isoform which in some cases is not similar enough for the mutations to be grounded onto it. In the FGFR3 case the recall was influenced by MutationFinder's inability to handle point mutations written in the form $p.wNm$, e.g. "p.Gly372Cys" as in [10], resulting in false negatives.

Our method is built upon the assumption that any article about mutations is about one single protein, whereas in reality or according to the retrieved COSMIC corpus, almost 14% of the documents describe mutations to a set of proteins instead of only one. As stated in the introduction, previous approaches for mutation grounding also have their shortcomings; [17] will fail whenever the position of the mutation mentioned in the text do not follow the numbering scheme of a database-stored sequence and [14], who handled the previous problem by building regular expressions containing wildtype residues of at least two mutations, will fail whenever one single mutation mention is incorrect or unrelated to the other ones. [15] also have the latter problem, which is now solved by our mutation grounding algorithm that tries to ground as many mutation mentions as possible instead of every mutation mention. [19] try to solve the problem with differences in numbering by using information from Swiss-Prot on e.g. post-translational cleavage. Since they do not use any advanced methods to associate proteins with mutations but instead rely on the keywords they used when retrieving the documents, they show very low precision (5 correctly grounded mutation mentions out of 222). This low performance can also be explained by the relatively high probability for a residue to fit any position of a sequence, i.e. $p = 1/20$. When also permitting for methionine cleavage and one post-translational cleavage the probability for an amino acid residue to match by chance is $p = 3/10$. As we use protein and organism names together with sets of mutations we can, using our algorithm, decrease the number of these false positives. [12] allow for alternative residue numbering for some proteins that present several isoforms or a signal sequence, but this is handled manually based on features extracted from Swiss-Prot for the protein families they considered. They show 87% precision which is in line with our results, but they have not measured recall, since they did not consider false negative point mutations not extracted from the documents.

In conclusion, automated reuse of mutation impact information from documents is now an achievable milestone, given the respectable performance of our grounding algorithm, albeit requiring testing on a larger corpus e.g. gold standard references provided by Swiss-Prot on protein families with complex numbering systems like Subtilases [22]. In combination with mutation impact

---

[7] The one included in the text version of Swiss-Prot.

extraction from sentences, the mutation grounding algorithm will facilitate the construction of unique datasets suitable as training material for predicting the impacts of genomic variations [4] and the extraction of genotype-phenotype relations [6]. To realize the construction of these datasets we will look into modularizing our current system into a series of web services that can be used by knowledge brokers in desired combinations suiting their individual needs.

Acknowledgements

This research was funded in part by the New Brunswick Innovation Foundation, New Brunswick, Canada. NSERC, Discovery Grant Program, Canada. And Quebec - New Brunswick University Co-operation in Advanced Education - Research Program, Government of New Brunswick, Canada. Funding was also provided in part by core funding from A*STAR (Agency for Science, Technology and Research), Singapore.

References

1. Baker, C.J.O., Witte, R.: Mutation Mining-A Prospector's Tale. Information Systems Frontiers 8, 47–57 (2006)
2. Bauher-Mehren, A., Furlong, L.I., Rautschka, M., Sanz, F.: From SNPs to pathways: integration of functional effect of sequence variations on models of cell signalling pathways. BMC Bioinformatics 10 (suppl. 8), S6 (2009)
3. Boeckmann, B., Bairoch, A., Apweiler, R., Blatter, M.C., Estreicher, A., Gasteiger, E., Martin, M.J., Michoud, K., O'Donovan, C., Phan, I., Pilbout, S., Schneider, M.: The Swiss-Prot Protein Knowledgebase and its supplement TrEMBL in 2003. Nucleic Acids Res. 31, 365–370 (2003)
4. Bromberg, Y., Rost, B.: SNAP: predict effect of non-synonymous polymorphisms on function. Nucleic Acids Res. 25(11), 3823–3835 (2007)
5. Caporaso, J.G., Baumgartner Jr., W.A., Randolph, D.A., Cohen, K.B., Hunter, L.: MutationFinder: a high-performance system for extracting point mutation mentions from text. Bioinformatics 23, 1862–1865 (2007)
6. Coulet, A., Shah, N., Hunter, L., Barral, C., Altman, R.B.: Extraction of Genotype-Phenotype-Drug Relationships from Text: From Entity Recognition to Bioinformatics Application. In: Pacific Symposium on Biocomputing, vol. 15, pp. 485–487 (2010)
7. Cunningham, H., Maynard, D., Bontcheva, K., Tablan, V.: GATE: A Framework And Graphical Development Environment For Robust NLP Tools And Applications. In: Proceedings of the 40th Anniversary Meeting of the Association for Computational Linguistics, ACL 2002 (2002)
8. Forbes, S.A., Bhamra, G., Bamford, S., Dawson, E., Kok, C., Clements, J., Menzies, A., Teague, J.W., Futreal, P.A., Stratton, M.R.: The Catalogue of Somatic Mutations in Cancer (COSMIC). Curr. Protoc. Hum. Genet. 57, 10.11.1–10.11.26 (2008)
9. Gabdoulline, R.R., Ulbrich, S., Richter, S., Wade, R.C.: ProSAT2–Protein Structure Annotation Server. Nucleic Acids Res. 34, W79–W83 (2006)

10. Hafner, C., Hartmann, A., Real, F.X., Hofstaedter, F., Landthaler, M., Vogt, T.: Spectrum of FGFR3 Mutations in Multiple Intraindividual Seborrheic Keratoses. Journal of Investigative Dermatology 27, 1883–1885 (2007)
11. Cotton, R.G.H., Horaitis, O.: The Challenge of Documenting Mutation Across the Genome: The Human Genome Variation Society Approach. Hum Mut. 23, 447–452 (2004)
12. Horn, F., Lau, A.L., Cohen, F.E.: Automated extraction of mutation data from the literature: application of MuteXt to G protein-coupled receptors and nuclear hormone receptors. Bioinformatics 20, 557–568 (2004)
13. Izarzugaza, J.M.G., Baresic, A., McMillan, L.E.M., Yeats, C., Clegg, A.B., Orengo, C.A., Martin, A.C.R., Valencia, A.: An integrated approach to the interpretation of Single Amino Acid Polymorphisms within the framework of CATH and Gene3D. BMC Bioinformatics 10(Suppl. 8), S5 (2009)
14. Kanagasabai, R., Choo, K.H., Ranganathan, S., Baker, C.J.O.: A Workflow for Mutation Extraction and Structure Annotation. J. Bioinformatics and Comp. Bio. 5(6), 1319–1337 (2007)
15. Krallinger, M., Izarzugaza, J.M.G., Rodriguez-Penagos, C., Valencia, A.: Extraction of human kinase mutations from literature, databases and genotyping studies. BMC Bioinformatics 10 (suppl. 8), S1 (2009)
16. Rebholz-Schuhmann, D., Marcel, S., Albert, S., Tolle, R., Casari, G., Kirsch, H.: Automatic extraction of mutations from Medline and cross-validation with OMIM. Nucleic Acids Res. 32, 135–142 (2004)
17. Winnenburg, R., Plake, C., Shroeder, M.: Improved mutation tagging with gene identifiers applied to membrane protein stability prediction. BMC Bioinformatics 10 (suppl. 8), S3 (2009)
18. Witte, R., Baker, C.J.O.: Towards a Systematic Evaluation of protein Mutation Extraction Systems. J. Bioinformatics and Comp. Bio. 5(6), 1339–1359 (2007)
19. Yip, Y.L., Lachenal, N., Pillet, V., Veuthey, A.-L.: Retrieving mutation-specific information for human proteins in UniProt/Swiss-Prot Knowledgebase. J. Bioinformatics and Comp. Bio. 5(6), 1215–1231 (2007)
20. Witte, R., Kappler, T.: Enhanced semantic access to the protein engineering literature using ontologies populated by text mining. International Journal of Bioinformatics Research and Applications 3(2), 389–413 (2007)
21. Erdogmus, M., Sezerman, U.: Application of automatic mutation-gene pair extraction to diseases. J. Bioinformatics and Comp. Bio. 5(6), 1261–1275 (2007)
22. Siezen, R.J., Leunissen, J.A.M.: Subtilases: the superfamily of subtilisin-like serine proteases. Protein Science 6(3), 501–523 (1997)

Handling Missing Features with Boosting Algorithms for Protein–Protein Interaction Prediction

Fabrizio Smeraldi[1,*], Michael Defoin-Platel[2], and Mansoor Saqi[2]

[1] School of Electronic Engineering and Computer Science,
Queen Mary University of London,
Mile End Road,
London UK-E14NS
fabri@dcs.qmul.ac.uk
http://www.dcs.qmul.ac.uk
[2] Biomathematics and Bioinformatics,
Rothamsted Research,
Harpenden, UK-AL52JQ
michael.defoin-platel@bbsrc.ac.uk,
mansoor.saqi@bbsrc.ac.uk
http://www.rothamsted.bbsrc.ac.uk

Abstract. Combining information from multiple heterogeneous data sources can aid prediction of protein-protein interaction. This information can be arranged into a feature vector for classification. However, missing values in the data can impact on the prediction accuracy. Boosting has emerged as a powerful tool for feature selection and classification. Bayesian methods have traditionally been used to cope with missing data, with boosting being applied to the output of Bayesian classifiers. We explore a variation of Adaboost that deals with the missing values at the level of the boosting algorithm itself, without the need for any density estimation step. Experiments on a publicly available PPI dataset suggest this overall simpler and mathematically coherent approach may be more accurate.

1   Introduction

One of the goals of systems biology is to understand the roles of proteins at various levels of biological organisation, from molecular function through to cellular and physiological function. The identification of networks of interacting proteins is a step to suggesting higher levels of organisation [4]. Experimental data for protein-protein interactions is available for a number of organisms in repositories such as Biogrid [5] and Intact [12]. However protein interaction datasets often contain many false positives [17,7,6] and are for most organisms largely incomplete. This has prompted the development of a number of machine learning

---

* Part of this work was done while F. Smeraldi was visiting Rothamsted Research.

P. Lambrix and G. Kemp (Eds.): DILS 2010, LNBI 6254, pp. 132–147, 2010.

approaches for the prediction of protein-protein interactions. Although information from sequence alone has been used [19,25], several approaches combine heterogeneous sources of data. Information from such sources can form components of a feature vector associated with a pair of proteins. Given a suitable Gold Standard comprising pairs of proteins that are known to interact and pairs known not to interact, a classifier can be constructed that attempts to discriminate interacting from non-interacting pairs. The features obtained from various data sources are widely heterogeneous and could include, for example, the extent to which the two proteins share similar patterns of co-expression, or the existence of sequence–similar proteins known to interact in another organism. The performance of the classifier can be assessed and the classifier can then be used to predict interactions between pairs of proteins for which no experimental information is yet available. Previous studies have constructed predicted interactomes for various organisms including yeast, *Arabidopsis* and human, using machine learning methods such as Bayes classifiers [11,15,24] and Support Vector Machines [3,25,13]. However, experimental measurements of each feature are usually available for different subsets of proteins pairs. Therefore, as more information is combined from multiple heterogeneous sources, the way that missing observations are treated becomes increasingly important. Arguably, this somewhat limits the available choices of classifiers, unless non-trivial pre-processing steps (typically Bayesian in nature) are applied. For instance, kernel methods offer no straightforward way to handle missing data, as the implicit mapping carried out by the kernel depends non-linearly on all components of the input vector (see for instance [20]); typical applications of ensemble methods such as AdaBoost [8] also rely on preprocessing.

In [11] a Naive Bayesian approach was used to predict protein-protein interactions in yeast by integrating a few genomic features. A larger number of features as well as a boosted classifier were used to assess the limits of data integration in [15]. Here we explore a variant of the Adaboost algorithm that is able to deal explicitly with missing feature values. Our approach matches or betters the results obtained by Bayesian preprocessing and is overall easier to implement. More importantly, it deals away with the arbitrariness inherent in density estimation and fits into the same solid theoretical framework as AdaBoost, especially with regard to convergence guarantees.

1.1   Dealing with Missing Features

Several strategies have been proposed for dealing with missing features (for an overview, see [20]). At least two typologies of approaches have been applied to the database we study (see Section 3). The first, most obvious strategy is simply to discard the examples for which data is missing. This approach has been investigated in [14], but complete information for each feature was only available for a small subset of the protein pairs.

Another classical technique that can be employed to deal with missing data is to complement the dataset with an estimate or a default value when a particular feature is missing. One way in which this has been done is by learning Naive

Bayes classifiers from (subsets of) the raw features [11]. Since Naive Bayes classifiers can fall back on the prior, missing data are no longer an issue (cf Section A.2); a more advanced classifier can then be applied to their output [15].

However, training a Naive Bayes classifier involves a density estimation step. Density estimation is an ill–posed and difficult problem [9]; the variety of the techniques available (Parzen windows, histograms, Expectation Maximisation to name a few) is indicative of the element of arbitrariness that this process entails. Indeed, some of the most successful classification algorithms such as Support Vector Machines [26] and Adaboost [8] owe their effectiveness at least in part to the fact that they avoid estimating densities for their input, but rather optimise the margin or a bound on the empirical error; from this point of view, introducing a preliminary density estimation step seems somewhat incongruous.

In this work we investigate an effective alternative to Naive Bayes classifiers for dealing with missing feature values when Adaboost [8] is used. Adaboost is an adaptive strategy for combining multiple binary classifiers with slightly better–than–random performance (the so-called *Weak Learners*) into a highly accurate *strong classifier*. The algorithm works by maintaining a list of weights over the training examples, so that "difficult" examples (that are misclassified by many Weak Learners) become more important over time. Adaboost iteratively chooses the optimal Weak Learner (WL) with respect to the weights, adds it to the strong classifier with an appropriate coefficient, and updates the weights of the training examples (for details, see Section 2). When each WL is a function of a single feature (for instance a simple threshold on the feature value, also known as a *decision stump*), Adaboost iterations essentially perform feature selection.

Training Bayesian classifiers on the raw feature values and then boosting them [15] amounts to dealing with missing data in the WLs, so that the missing values are hidden from Adaboost. The alternative we here explore is to use WLs that *abstain* (i.e. do not return a decision) on missing data, and let the boosting algorithm deal with the problem.

Although variants of AdaBoost for WLs that abstain have been introduced early on [23], they have not been as widely applied as the standard algorithm; specifically, to the best of our knowledge, they have not yet been applied to PPI data. One of the reasons for this may be that these algorithms have been introduced in the slightly different context of confidence-rated predictions, i.e. under the assumption that the WLs provide a graded output instead than a binary decision. Also, while a considerable computational simplification of standard Adaboost has made the implementation of the algorithm straightforward [2], to the best of our knowledge this has not yet been generalised to the case of WLs that abstain.

In this paper we introduce in detail the simplified version of the algorithm for classifiers that abstain and we apply it to a widely investigated set of features for PPI in yeast [15]. Our results show that avoiding density estimation and dealing with missing features at a late stage may indeed be the better option, at least when Boosting algorithms are used.

Table 1. Simplified version of AdaBoost for weak learners that abstain (Ada-ABS). This algorithm is equivalent to the one introduced in [23] and reduces to standard Adaboost in the case of weak learners that never abstain, as is the case with Naive Bayes classifiers.

Adaboost for weak learners that abstain (simplified):

Input:

1. Labelled training vectors $\{(\boldsymbol{x}_i, y_i)\}$, with $y_i \in \{-1, +1\}$.
2. Weak learners $\{h_j(x) : \mathsf{X} \to \{-1, 0, +1\}\}$.

Initialise the weights $d_{1,i} = \frac{1}{m}$.

For $t = 1, \ldots, T$:

1. Train each weak learner $h_j(\boldsymbol{x})$. Using the current distribution of weights $d_{t,i}$, compute the total weight of the training examples on which it abstains (W_a), that it classifies correctly (W_c) and that it misclassifies (W_m).
2. Select the weak learner h_t that minimises $Z = W_a + 2\sqrt{W_c W_m}$. Check that $Z < 1$; otherwise quit.
3. Compute $\alpha_t = \frac{1}{2}\log(W_c/W_m)$ and update the weights:

$$d_{t+1,i} = \begin{cases} d_{t,i}/(W_a\sqrt{W_m/W_c} + 2W_m) & \text{if } h_t \text{ misclassifies } \boldsymbol{x}, \\ d_{t,i}/(W_a\sqrt{W_c/W_m} + 2W_c) & \text{if } h_t \text{ classifies } \boldsymbol{x} \text{ correctly}, \\ d_{t,i}/Z & \text{if } h_t \text{ abstains on } \boldsymbol{x}. \end{cases}$$

Output the strong classifier: $H(\boldsymbol{x}) = \text{sign}\left(\sum_{t=1}^{T} \alpha_t h_t(\boldsymbol{x})\right)$

---

2   Boosting Weak Learners That Abstain

We use the AdaBoost algorithm for classifiers that abstain given in [23], since the choice of parameters in this algorithm guarantees the tightest bound on the training error. However, instead of following the traditional formulation of Adaboost, we extend to the case of classifiers that abstain the simplification presented in [2]. As shown in [16], such simplification can be seen as a direct solution of the dual formulation of the Adaboost minimisation problem. Extension to the case of WLs that abstain is straightforward, and leads to the algorithm outlined in Table 1 (henceforth Ada-ABS). We emphasise that this simplified version computes, step by step, the same weights and the same final decision function as specified in [23]. The main advantage is its simplicity; by comparison, in the original formulation of Adaboost [8] the weight update procedure is

$$d_{t+1,i} = \frac{1}{Z_t} d_{t,i} \exp\left(-\alpha_t y_i h_t(\boldsymbol{x}_i)\right), \tag{1}$$

with

$$Z_t = \sum_i d_{t,i} \exp\left(-\alpha_t y_i h_t(\boldsymbol{x}_i)\right) \tag{2}$$

The algorithm for classifiers that abstain is very similar to classical Adaboost, with the following main differences:

1. the value of α is $\log(W_c/W_m)/2$ (different variants of the algorithm specify other choices for α);
2. at each iteration the weak learner that minimises Z is chosen, as opposed to the learner that minimises the weighted classification error W_c.

The reason for points 2. and 3. above is that, at each iteration, Z multiplies a bound on the training error; the choice of α minimises Z given W_c and W_m. When $W_a = 0$, i.e. the weak learner never abstains, this is equivalent to the standard choice $\alpha = 1/2\log((1 - W_m)/W_m)$. It is easy to see that if no weak learners abstain also the weight update equations, and hence the entire algorithm, revert to standard Adaboost.

Notice that, since $Z = W_a + 2\sqrt{W_c W_m}$, weak learners that abstain on a large number of training examples (or on training examples carrying a large weight) are penalised irrespective of their performance on the examples on which they do provide a decision. Also, since $Z_t < 1$, training examples on which a weak learner abstains see their weight increased for the following iteration. Therefore, the algorithm will eventually select one or more weak learners to classify each of the training examples.

3   Dataset

In this study we use a Gold Standard dataset[1] for the prediction of protein-protein interactions as described in [11] and [15]. This is based on the MIPS (Munich Information Centre for Protein Sequences) hand curated catalogue. We report experimental results over a subset of 3161 proteins, as already described in [15]. This subset contains 2,711,441 interacting or non-interacting protein pairs, for which at least one of the 16 genomic features defined in Table 2 is available. For a complete definition of the features and of their relationship to protein-protein interactions see [15]. In the whole dataset, a sample consists of two protein names, 16 genomic features and a label that specifies if the proteins interact or not. The distribution of the features is shown in Figure 1. The dataset has a large number of missing features, even after discarding from the dataset the samples for which no data at all is available; the percentage of missing data by feature is shown in Table 2.

4   Results

We compare the performance of the three *strong classifiers* obtained by boosting three different sets of WLs. These are implemented as threshold-based classifiers

---

[1] Available online at http://networks.gersteinlab.org/intint

Table 2. Percentage of missing values by feature over the 8,250 interacting protein pairs, 2,703,191 non-interacting pairs (and 2,711,441 pairs total) listed in the Gold Standard after removing the pairs for which no feature values at all are given. Also shown is a rank of the feature according to information gain [18] for the original dataset and a dataset complemented by the mean.

Features	Description	Missing data (%)			IG Rank	
		Positive	Negative	Total	Orig.	Compl.
F1	mRNA Co-expression	7.71	1.03	1.05	7	4
F2	MIPS Functional Similarity	2.41	51.41	51.26	6	1
F3	GO Functional Similarity	8.85	76.06	75.86	5	2
F4	Co-Essentiality	73.85	78.78	78.76	8	8
F5	Absolute mRNA Expression	5.62	0.27	0.28	9	5
F6	Marginal Essentiality	6.21	4.25	4.26	10	7
F7	Absolute Protein Abundance	37.07	43.97	43.95	11	11
F8	Co-regulation	52.15	97.79	97.65	16	16
F9	Phylogenetic Profiles	88.92	99.03	99.00	4	6
F10	Gene Neighbourhood	96.22	99.96	99.95	3	10
F11	Rosetta Stone	98.63	99.95	99.95	2	9
F12	Synthetic Lethality	98.75	99.97	99.97	15	15
F13	Gene Cluster	99.98	99.98	99.98	14	14
F14	Threading Scores	98.75	99.96	99.95	13	13
F15	Co-evolution Scores	99.98	99.99	99.99	12	12
F16	Interologs in other Organism	54.65	99.85	99.71	1	3

(*decision stumps*), a common choice for WLs (see Section A.1). The WLs are trained (*i*) on the log odds of interaction for a particular feature (Bayes WLs), (*ii*) on the raw feature values (WL-Abs) or (*iii*) on the raw feature values augmented by the Naive Bayes score (Section A.2), considered as an additional feature (WL-Abs+Naive Bayes). WLs trained on the raw feature values abstain (return a value of zero) when the feature value is missing. We use the Ada-ABS algorithm outlined in Table 1. It should be noted that, as detailed in Section 2, Ada-ABS reverts to standard Adaboost when the weak learners do not in fact abstain.

A Naive Bayes Classifier (NBC) trained on all features is used to give a baseline performance. A common objection to boosting approaches based on decision stumps is that correlations between feature values or higher order statistical moments that might be exploited, for instance, by a Support Vector Machine are treated very poorly, as each WL bases its binary decision on a single feature.

As a coarse estimate of this limitation, we augment the set of WLs that abstain with a decision stump based on the posterior estimate provided by the baseline NBC. As we will see, when the percentage of missing data is limited this combination of WLs that abstain with an NBC improves performance well beyond the level achieved by the classical approach of applying Bayesian estimation to the single features separately and then boosting them. As the percentage of missing data increases our results suggest that the NBC becomes less informative; weak learners that abstain still provide optimal results in this case.

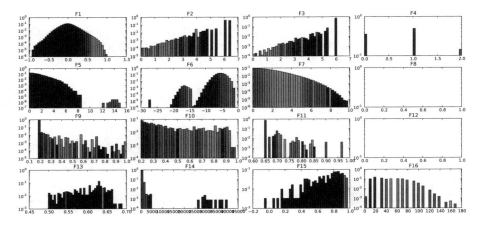

Fig. 1. Distributions of the feature values (on a log scale). Colour indicates the fraction of interacting protein pairs in each bin. Blue indicates that the bin mainly corresponds to non-interacting pairs; red bins predominantly account for interacting pairs.

4.1   Classification Accuracy

We perform a series of four-fold cross-validation experiments on the dataset described in Section 3 with an increasing number n of boosting iterations, using different subsets of the features listed in Table 2. A schematic of the results is given in Table 3.

Figure 2 shows the Equal Error Rate averaged across the four cross-validation runs as a function of n, when all the 16 features are used for training.

The minimum error rate when boosting Bayes classifiers is in this case achieved after 207 iterations (average EER=3.2%). For WLs that abstain, a slightly lower EER is obtained after 51 iterations (average EER=3.0%). If we consider the estimate of the NBC as a new feature and boost it with Ada-ABS together with the WLs that abstain, the average EER is lowered significantly to 2.0% (at 293

Table 3. Summary of the EER for different choices of classifiers and of features, averaged over 4 cross-validation rounds. The database and the first two algorithms listed were explored in [11,15]. Bold values show the minimal EER by set of features. Boosting WLs that abstain is always better than both Naive Bayes and the boosted Bayesian WLs. Complementing the WLs that abstain with the NB classifier can improve performance when multiple observations with low percentages of missing data are used.

EER (%)	All features	F1–F4	F5–F16	w/o F2,F3
Naive Bayes Class.	5.0%	5.3%	27.3%	12.2%
Bayes WLs	3.2%	3.5%	18.6%	13.0%
WLs that Abstain	3.0%	3.3%	**17.0%**	**11.7%**
WL-Abs + NBC	**2.0%**	**2.5%**	**17.0%**	12.2%

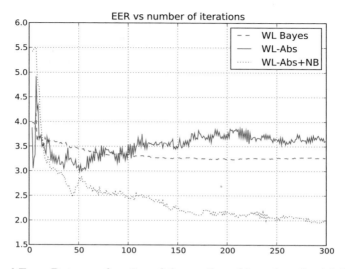

Fig. 2. Equal Error Rate as a function of the number of iterations for AdaBoost using Bayes WLs, WLs that abstain and the combination of WL-Abs and the Naive Bayes Classifier trained on all features.

iterations). This is more that 30% better than boosting the Bayes WLs and less than half the average EER of the baseline Naive Bayes Classifier (which is 5.0%).

The ROCs corresponding to the optimal average EERs listed above are displayed in Figure 3, for all four cross-validation rounds separately. As can be seen, all boosting approaches widely improve on the Naive Bayes. Weak Learners that abstain yield slightly better performance than boosted Bayes WLs, with a more marked advantage in the high sensitivity part of the curve. The best ROC curves are obtained by adding the Naive Bayes Classifier as a weak learner, arguably because this captures the correlation between the features — a point that is confirmed by our results in Section 4.3 below.

In predictive usage, one would need to set the number of iterations using a validation set and extrapolate. This applies both to Adaboost with Bayes WLs and to the algorithms we introduce. Convergence and generalisation ability of Adaboost as a function of the number of iterations is a widely investigated topic [22,21], that is largely beyond the scope of this work. Empirically from Figure 2 we notice that, after the decision function stabilises in the first 30 iterations, the boosted classifiers are fairly insensitive to the choice of the number of iterations: the average EER oscillates in a band narrower than 1% as n is increased from 30 to 300. However, Ada-ABS has a slight tendency to over-fit after about 60 iterations, while Bayesian WLs are more stable. This is likely due to the fact that density estimation with a histogram (Section A.2) effectively reduces the choices for the weak learners by limiting the number of different thresholds available for each feature to the number of bins in the histogram.

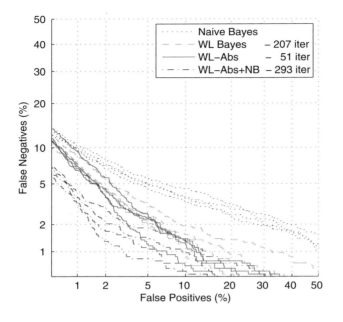

Fig. 3. Superposed ROC curves for all 4 cross validation rounds. The curves displayed correspond to Bayes WLs (207 iterations), WLs that abstain (51 iterations) and a combination of WL-Abs and the Naive Bayes classifier (293 iterations).

In turn, this seems to regularise the decision function, albeit at the expense of classification accuracy.

However, it should be noted that while the curves in Figure 2 tell the entire story for WL-Abs, of which the number of iterations is the only parameter, Adaboost with Bayesian (weak) classifiers includes in addition many more parameters hidden in the density estimation step; this should be kept into account when assessing the relative stability of the algorithms.

4.2   Structure of the Decision Function

In Table 4 we list the features that contribute to the decision function for each of the ROC curves in Figure 3. Features are listed in the order in which the corresponding WLs are first selected by the boosting algorithm; the number of times a particular feature is chosen (possibly with different thresholds) is indicated in parentheses.

As can be seen, the order in which the features are chosen is fairly stable for each set of WLs, with variations between the cross-validation rounds occurring for higher numbers of iterations. By comparison with Table 2 we see that the first few features are chosen among the most discriminative features and among those with the lowest percentage of missing values. Feature F5, the first feature when

Table 4. Features used by the WLs appearing in the decision functions that yield the ROC curves in Figure 3. WLs are listed in order of first appearance. The number of times each feature is chosen is given in parentheses. Multiple comma–separated entries correspond to variations across the four cross-validation runs. NB indicates the WL corresponding to the Naive Bayes Classifier.

Classifier	Weak learners (occurrences)
WLs that Abstain (51 iter.)	F5(3,5,6) – F2(15,18,19) – F1(7,9,11) – F3(4,5,7) – F6(5,9,7,11) – F16(2) – F4(2),F7(1,2) – F8,12(1), F4(2) – F8,12(1) – F7,8(0,1) – F14(0,1)
Bayes WLs (207 iter.)	F16(31,33) – F2(30,37,38) – F3(25,27,30) – F1(19,23,27,34) – F5(17,18,25,34) – F4(6,7,9) – F6(20,24,29,30) – F8(4,5) – F7(3,4,6,7) – F12(9,13-15) – F14(0,2,6,7) – F11(0,3) – F9(0,1,2)
WLs that Abstain plus Naive Bayes (293 iter.)	NB(64,78,80) – F1(10,16,25) – F2(91,108,109,112) – F5(11,18,19), F16(2) – F5(28),F16(1,2) – F3(37,40-42) – F4(2),F6(21,25,30) – F4(2,3),F6(18) – F8(2),F12(1) – F8(1,2),F12(1) – F14(2,3),F7(8) – F7(5,10),F9(2),F11(1) – F7(1),F9(1,2),F11(1) – F11(1),F14(0,1)

using Weak Learners that Abstain, has the lowest total percentage of missing data, while F16 (the first Bayesian WL) has the third highest information gain on the complemented dataset (and the highest on the original dataset).

Probably as an effect of the unbalance between positive and negative training examples, overlap with the positive training data seems to be more important than the coverage of the set of non-interacting pairs. Feature 16 for instance has 99.71% missing data but covers about 45% of the positive training set. Similarly Feature 3 has 75.86% of missing data overall, but only just short of 9% of the positive training data are not covered. Conversely, features 10, 13 and 15 do not appear at all in the optimal decision functions (except insofar as they contribute to the NBC). As shown in Table 2, these features are unavailable for at least 99.95% of the Gold Standard data, with very limited coverage of both interacting and non-interacting pairs.

Table 4 provides a picking order, but it does not specify the weight of each feature in the decision function nor the dependence on the feature value. These data are displayed in Figure 4 for the decision function resulting from 51 iterations of Ada-ABS on WLs that abstain.

More specifically, with reference to the strong decision function $H(x)$ defined in Table 1, for each feature j the figure shows

$$H_j(x) = \sum_{t \in S_j} \alpha_t h_t(x_j) \tag{3}$$

where S_j is the set of iterations in which a weak learner that is a function of feature j has been chosen . As can be seen, the functions H_J corresponding to the 11 distinct features all have similar range, suggesting that they all contribute significantly to the final decision. This confirms that the classifier can effectively extract information from features with high percentages of missing data.

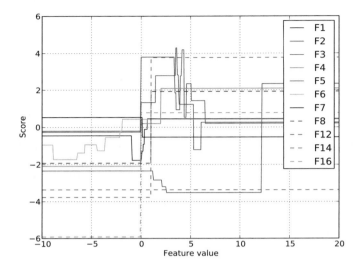

Fig. 4. Contribution of the single features to the decision function for Ada-ABS, 51 iterations. F6 has an additional threshold at -27.63, F14 at +287.00, and F16 at +50.00.

Figure 4 also shows how boosting can give different weights to specific intervals of feature values by combining simple step-wise weak learners with different thresholds. The type of these intervals (open or close) also depends on the parity of the number of weak learners for the specific feature. As the first row of Table 4 shows, a few features have a marked preference for an odd or an even number of occurrences. This explains the jagged appearance of the EER curves for WLs that abstain in Figure 2 .

4.3   Exploring Subsets of Features

According to [15], near-optimal performance should be achieved in the case of boosting Bayes WLs even when only the first four features are used (mRNA co-expression, MIPS functional similarity, GO functional similarity and Co-essentiality). We test this by limiting the boosting algorithm to select among weak learners trained on the first four features only (both for Bayes WLs and WLs that abstain). The resulting average EERs are reported in the third column of Table 3. Indeed, for both the baseline Naive Bayes Classifier and boosted Bayes WLs, the average EERs increase by less than 10%. The lowest average EER of 2.5%, however, is obtained using a combination of WLs that abstain with a WL that thresholds the NBC; this is less than half the EER of the baseline NBC (5.3%), and actually improves on both the NBC and boosted Bayes WLs even when those are allowed to use all features (see the second column of Table 3). It should be noted that, when we allow our algorithm to use all features, the EER is reduced by a further 25%. This suggests that F1 to F4 do not

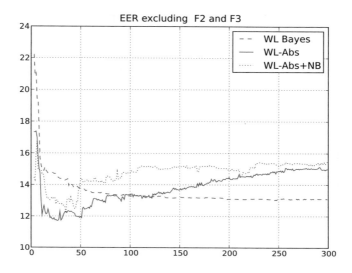

Fig. 5. Equal Error Rate as a function of the number of iterations when boosting over all features except for F2 and F3, that are based on functional similarity.

actually capture all the information available, and that our approach is better at extracting information from the remaining features (F5 to F16).

To further test the behaviour of our algorithms on features that mostly have high percentages of missing data and have previously been found to be less informative, we perform cross-validation tests on the database using features F5 to F16 only. Average EERs are displayed in the fourth column of Table 3. As can be seen, WLs that abstain outperform the other approaches. Adding a WL based on the Naive Bayes classifier does not in this case improve accuracy; this is, in our view, a consequence of the high number of missing feature values and of the consequent overall poor performance of the NBC itself.

Following [15] we also explore the effect of removing two strong features, namely F2 and F3, that are based on functional similarity and dominate the prediction performance. Since the accurate assignment of a functional category to a protein generally involves a manual curation step, prediction performance without these features may also better reflect a more general application case.

In Figure 5 we report the average EER as a function of the number of iterations after removing features F2 and F3 from the database. The minimum average EER for boosted Bayes WLs is 13.0%, while for WLs that abstain it is 11.7% (see Table 3, last column). These values are achieved after 246 and 27 iterations respectively. The average EER of the baseline Naive Bayes Classifier is 12.2%; adding its output as a WL to the WLs that abstain does not in this case improve accuracy. The corresponding ROC curves for all four cross-validation round and the best choice of the number of iterations are displayed in Figure 6.

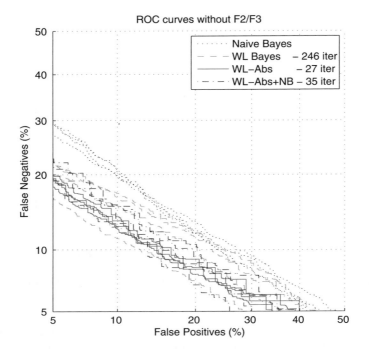

Fig. 6. ROC curves for Naive Bayes classifiers (33 iterations) and Raw Features (27 iterations). Curves for the four cross-validation rounds are superposed. Note how boosting the Raw Features leads to higher precision as well as increased stability.

5   Conclusions

The integration of multiple heterogeneous data sources is important in systems biology as it has the potential to impact on prediction. In order to exploit information that may be available only on parts of the dataset, effective strategies for handling missing data are required.

We explore the use of a variant of Adaboost (originally derived in the context of confidence–rated predictions) that allows the weak learners to "abstain", i.e. not to return a decision whenever a feature value is missing. We introduce an exact simplified mathematical formulation of this algorithm along the lines of an existing simplification of AdaBoost. The simplified algorithm is hardly more complicated to implement than Adaboost, and reverts to standard Adaboost when no data are missing.

We test our algorithm over different subsets of features from a widely used database of yeast PPI data, that includes measurements with high numbers of missing features. Experimental results show that our approach can handle large percentages of missing data effectively, consistently outperforming the common alternative of boosting single-feature Bayesian classifiers. Besides being overall easier to implement, our approach deals away with the theoretical and practical complications of density estimation.

Our results also indicate that a Naive Bayes classifier trained on all features, although overall far less effective on its own, may capture correlations between the features that escape an approach based on decision stumps; when this happen, this information can be integrated in the boosting framework by considering the Naive Bayes classifier as an additional weak learner. In future work we propose to investigate the use of other choices for the weak learners that may better account for dependency between the features.

References

1. Azuaje, F., Dopazo, J.: Data Analysis and Visualization in Genomics and Proteomics. John Wiley & Sons, Chichester (2005)
2. Bauer, E., Kohavi, R.: An empirical comparison of voting classification algorithms: bagging, boosting, and variants. Machine Learning 36(1-2), 105–139 (1999)
3. Ben-Hur, A., Noble, W.S.: Kernel methods for predicting protein-protein interactions. Bioinformatics 21(Suppl. 1), i38–i46 (2005)
4. Bork, P., Jensen, L.J., von Mering, C., Ramani, A.K., Lee, I., Marcotte, E.M.: Protein interaction networks from yeast to human. Curr. Opin. Struct. Biol. 14(3), 292–299 (2004)
5. Breitkreutz, B.J., Stark, C., Reguly, T., Boucher, L., Breitkreutz, A., Livstone, M., Oughtred, R., Lackner, D.H., Bhler, J., Wood, V., Dolinski, K., Tyers, M.: The bioGRID interaction database: 2008 update. Nucleic Acids Res. 36(Database issue), D637–D640 (2008)
6. Deane, C.M., Salwiski, L., Xenarios, I., Eisenberg, D.: Protein interactions: two methods for assessment of the reliability of high throughput observations. Mol. Cell. Proteomics 1(5), 349–356 (2002)
7. Edwards, A.M., Kus, B., Jansen, R., Greenbaum, D., Greenblatt, J., Gerstein, M.: Bridging structural biology and genomics: assessing protein interaction data with known complexes. Trends Genet. 18(10), 529–536 (2002)
8. Freund, Y., Schapire, R.: A decision-theoretic generalization of on-line learning and an application to boosting. Journal of Computer and System Science 55(1) (1997)
9. Hastie, T., Tibshirani, R., Friedman, J.H.: The Elements of Statistical Learning. Springer, Heidelberg (2001)
10. Jansen, R., Greenbaum, D., Gerstein, M.: Relating whole-genome expression data with protein-protein interactions. Genome Res. 12(1), 37–46 (2002)
11. Jansen, R., Yu, H., Greenbaum, D., Kluger, Y., Krogan, N.J., Chung, S., Emili, A., Snyder, M., Greenblatt, J.F., Gerstein, M.: A Bayesian networks approach for predicting protein-protein interactions from genomic data. Science 302(5644), 449–453 (2003)
12. Kerrien, S., Alam-Faruque, Y., Aranda, B., Bancarz, I., Bridge, A., Derow, C., Dimmer, E., Feuermann, M., Friedrichsen, A., Huntley, R., Kohler, C., Khadake, J., Leroy, C., Liban, A., Lieftink, C., Montecchi-Palazzi, L., Orchard, S., Risse, J., Robbe, K., Roechert, B., Thorneycroft, D., Zhang, Y., Apweiler, R., Hermjakob, H.: Intact–open source resource for molecular interaction data. Nucleic Acids Res. 35(Database issue), D561–D565 (2007)

13. Lin, M., Hu, B., Chen, L., Sun, P., Fan, Y., Wu, P., Chen, X.: Computational identification of potential molecular interactions in Arabidopsis. Plant Physiol. 151(1), 34–46 (2009)
14. Lin, N., Wu, B., Jansen, R., Gerstein, M., Zhao, H.: Information assessment on predicting protein-protein interactions. BMC Bioinformatics 5, 154 (2004)
15. Lu, L.J., Xia, Y., Paccanaro, A., Yu, H., Gerstein, M.: Assessing the limits of genomic data integration for predicting protein networks. Genome Res. 15(7), 945–953 (2005)
16. Malacaria, P., Smeraldi, F.: On Adaboost and optimal betting strategies. In: Proceedings of the 5th international conference on data mining (dmin/worldcomp), July 2009, pp. 326–332. CSREA Press (2009)
17. von Mering, C., Krause, R., Snel, B., Cornell, M., Oliver, S.G., Fields, S., Bork, P.: Comparative assessment of large-scale data sets of protein-protein interactions. Nature 417(6887), 399–403 (2002)
18. Michalski, R.S., Carbonell, J.G., Mitchell, T.M.: Machine Learning: An Artificial Intelligence Approach. Tioga Publishing Company (1983)
19. Najafabadi, H.S., Salavati, R.: Sequence–based prediction of protein–protein interaction by means of codon usage. Genome Biology 9(5) (2008)
20. Pelckmans, K., Brabanter, J.D., Suykens, J.A.K., Moor, B.D.: Handling missing values in support vector machine classifiers. Neural Networks 18, 684–692 (2005)
21. Rätsch, G., Warmuth, M.: Efficient margin maximizing with boosting. Journal of Machine Learning Research 6, 2131–2152 (2005)
22. Rudin, C., Schapire, R.E., Daubechies, I.: On the dynamics of boosting. In: Advances in Neural Information Processing Systems, vol. 16 (2004)
23. Schapire, R., Singer, Y.: Improved boosting algorithms using confidence-rated predictions. Machine Learning 37(3) (1999)
24. Scott, M.S., Barton, G.J.: Probabilistic prediction and ranking of human protein-protein interactions. BMC Bioinformatics 8, 239 (2007)
25. Shen, J., Zhang, J., Luo, X., Zhu, W., Yu, K., Chen, K., Li, Y., Jiang, H.: Predicting protein-protein interactions based only on sequences information. Proc. Natl. Acad. Sci. U.S.A. 104(11), 4337–4341 (2007)
26. Vapnik, V.N.: The nature of statistical learning theory. Springer, Heidelberg (1995)

A   Appendix

A.1   Weak Learners That Abstain

We use decision stumps that abstain to deal with missing feature values in the dataset. More in detail, given training vectors x with missing components we define a weak learner for each of the N features (components) in the following way:

$$h_j(x) = \begin{cases} +p_j & \text{if } x_j \geq \tau_j \\ -p_j & \text{if } x_j < \tau_j \\ 0 & \text{if } x_j \text{ is missing.} \end{cases} \tag{4}$$

The "polarity" constant $p_j \in \{-1, +1\}$ allows us to cater for features representing similarities and dissimilarities in the same framework. Both p_j and τ_j are optimised at each iteration to minimise Z as defined in Table 1.

A.2   Naive Bayes Classifiers

A Naive Bayes Classifier is based on the Bayes Rule and on the assumption that the features are conditionally independent given the class. It assigns the most probable label value $\hat{y}$ to the vector $\boldsymbol{x} = \{x_1, \ldots, x_N\}$ according to:

$$\hat{y} = \text{sign} \left(\frac{P(+1)}{P(-1)} \prod_j^N \frac{P(x_j| +1)}{P(x_j| -1)} - \tau \right) \tag{5}$$

where $P(y)$ is the prior probability of class y, $P(x_j|y)$ are the class-conditional densities and τ is a threshold used to set the operation point of the classifier.

NBCs provide a straightforward way of dealing with missing feature values: when a given feature x_j is missing the corresponding likelihood ratio $P(x_j| +1)/P(x_j| -1)$ in Equation 5 is set to 1. In the limit case that all the features of $\boldsymbol{x}$ are missing, the classifier can still return a prediction relying on the *a-priori* odds $P(+1)/P(-1)$.

Arguably the most critical part in Naive Bayes classification is the computation of robust estimates for the densities $P(y)$ and $P(x_j|y)$. In this study, the prior probabilities $P(y)$ are estimated as the proportion of each class in the training set. The conditional probabilities $P(x_j|y)$ are more difficult to estimate, mainly because many of the the x_j are continuous features. The seminal work of [10] has shown that a simple discretization technique by histograms can be used for this dataset. When known (see for example [1]), the numbers and ranges of the bins used in former studies have been employed in this work; otherwise, similar binning strategies have been used.

In order to obtain smooth estimates, the m-estimate method was used with $m = 2$ to compute the probabilities from frequencies of labels according to

$$P(x_j = b|y) = \frac{\text{count}(b, y) + mP(y)}{\text{count}(b) + m} \tag{6}$$

where $\text{count}(b, y)$ is number of samples with label y in bin b, $\text{count}(b)$ is the total number of samples in bin b and $P(y)$ is the prior probability of label y.

In this study, the classifier defined by Equation 5 is referred to as the Naive Bayes Classifier. The single–feature Bayes WLs are obtained by thresholding the likelihood ratio $P(x_j| +1)/P(x_j| -1)$ for each individual feature $x_j \in \boldsymbol{x}$.

Instance Discovery and Schema Matching with Applications to Biological Deep Web Data Integration

Tantan Liu, Fan Wang, and Gagan Agrawal

Department of Computer Science and Engineering
Ohio State University, Columbus OH 43210
{liut,wangfa,agrawal}@cse.ohio-state.edu

Abstract. This paper presents data mining-based techniques for enabling data integration across deep web data sources. We target query processing across inter-dependent data sources. Thus, besides input-input and output-output matching of attributes, we also need to consider input-output matching. We develop data mining techniques for discovering the instances for querying deep web data sources from the information provided by the *query interfaces themselves*, as well as from the obtained output pages of the related data sources, by query probing using *dynamically* identified input instances. Then, using a hierarchical representation of schemas and by applying clustering techniques, we are able to generate schema matches. We show the effectiveness of our technique while integrating 24 query interfaces.

1   Introduction

One of the critical problems that most scientific (and commercial) communities face is the explosion of data and data sources. As a result, integration has become an important phase in biology research process. Much effort has been put into developing general information integration methods, as well as integration tools specific to particular domains [6]. Data mining can help reduce the efforts in developing integrated systems. For example, many techniques have been developed for automatic wrapper generation for HTML data sources [2]. However, the introduction of new types of data sources and the desire for new integrated search methods is leading to new scenarios where novel data mining techniques are needed. This paper describes a case study in developing data mining techniques for a novel integration problem.

1.1   Application Context

The data mining application we have studied arises in the context of integrating *deep web* data sources and providing *keyword search* functionality. Deep web is a term coined to describe the contents that are stored in the databases and can be retrieved over the internet by querying through HTML forms. The deep web is estimated to be several orders of magnitude larger than the *surface web*,

P. Lambrix and G. Kemp (Eds.): DILS 2010, LNBI 6254, pp. 148–163, 2010.

where the HTML pages are static, and therefore, can be easily indexed by search engines [5].

The deep web contains millions of HTML forms, and covers a spectrum of information that is becoming not only indispensable for the normal functioning of today's human society, but is also becoming a vital facilitator of scientific research. Important scientific data is being shared through the deep web. One of the most successful examples is the publication of biological data through hundreds of the deep web sites globally, which is enabling biological and medical researchers to share and download much valuable information by simply filling online web forms.

Standard search engines like Google are typically not able to crawl the structured contents hidden inside the deep web-sites. A number of efforts have been attempting to build deep web systems that can integrate both the query interface and the query results (structured data) of the deep web-sites within a specific domain [14]. However, a challenge that has not been addressed arises because of inter-dependence between the data sources. Let us consider a motivating scenario in bioinformatics. Suppose we are interested in Single Nucleotide Polymorphisms (SNPs), which are particularly promising for explaining the genetic contribution to complex diseases [1]. Over seven million Single Nucleotide Polymorphisms (SNPs) and their relevant information have been reported in a number of public databases (deep web data sources). However, much information that biological researchers are interested in requires a search across multiple different deep web databases. No single database can provide all user requested information, and the output of some databases need to be the input for querying another database. Specifically, consider a query that asks for the amino acids occurring at the corresponding position in the orthologous gene of non-human mammals with respect to a particular gene, such as ERCC6 [18]. There is no database which takes gene name ERCC6 as input, and outputs the corresponding amino acids in the orthologous gene of non-human mammals. Instead, one needs to execute the following query plan. We first need to use gene name ERCC6 as input to query on an SNP database such as dbSNP to find all non-synonymous SNPs. Then, we, still taking ERCC6 as input, use a gene database, such as Entrez Gene, to obtain the encoded proteins in human species and other orthologous species. After that, using the proteins obtained from Entrez Gene, we search a sequence database to find the sequences. Finally, we use the sequences and the SNP obtained from dbSNP as input to do an alignment using an alignment database such as Entrez BLAST. Clearly, given a keyword query from a user, automatically determining a query plan as complicated as this one is a difficult problem.

Besides the need for effective query planning [18], another major problem is reducing the effort required in integration. This challenge is significant because of the rapidly increasing number of data sources. For example, there are more than 1000 online databases in the biological domain and the number is still increasing rapidly every year [3].

Each deep web data source may have multiple input query interfaces and each input interface corresponds to a unique input schema. Thus, we have the *input* attributes, which are required in the input interface, and the *output* attributes, which are the attributes returned by querying the corresponding input interface. We assume all their attributes belong to a universal set of attributes. For two deep web sources R_1 and R_2, we define that R_2 is *dependent* on R_1 if R_1 can provide any of the input attributes for R_2. An example of such dependence between several biological data sources is shown in Figure 1.

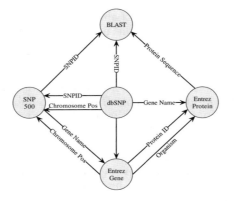

Fig. 1. Dependence Relations between Five Data Sources

1.2 Problem Definition and Contributions

The goal of schema matching on deep web data sources is to find the semantic correspondence between data sources. As we mentioned earlier, a deep web data source has two types of schemas: input schema and output schema. Correspondingly, there are three types of schema matching on deep web data source: 1) input-input schema matching across different data sources, 2) output-output schema matching, and 3) input-output schema matching. Among these, input schema matching[9,10,22] has been well studied in past five years. It tries to find the semantic correspondence between query interfaces of deep web sources and aims at providing a unified interface to the user. In contrast, our goal is discovering schema matching between output schema and input schema across different web data sources. For example, in the context of Figure 1, the goal will be to build a graph of the type shown here automatically or semi-automatically.

In this paper, we develop a series of data mining techniques for input-output schema matching across biological web data sources. The main aspect of our approach is a technique for *Discovering Instances*. We show that the instances for querying deep web data sources can be discovered from the information provided by the *query interfaces themselves*, as well as the obtained output pages of related data sources by query probing using *dynamically* identified input instances. Though instance based schema matching has been studied previously [21,20], our approach is distinct and novel. Particularly, WebIQ [21] discovers the instances of an attribute using the knowledge learned from surface web (Google search), and Wang *et al.* [20] use a predefined or *static* set of input instance to perform query probing.

Two other important aspects of our work are as follows. We use a hierarchical model to capture the output schemas in biological data sources. The hierarchical model can help reveal the underlying relations among the attributes. Also, we identify the mappings between the attributes based on a clustering approach.

The clustering approach has the *bridging effect* [15], i.e. the semantic similarity between two attributes can be effectively discovered if they are both semantically similar to a third attribute.

2   Basic Formulation

This section describes the models we use to capture input and output schemas, and the similarity function we use. The main algorithm is presented in the next Section. We use a running example to explain the main ideas from this and the next Section. This example is presented in Figure 2.

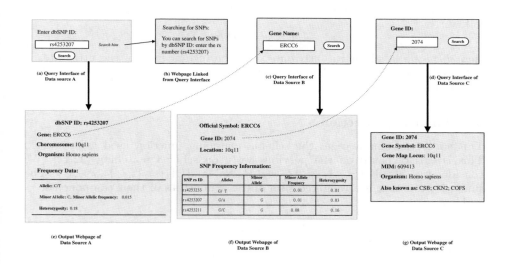

Fig. 2. Query Interfaces and Output Webpages of Data Sources

2.1   Modeling of Input Schemas and Output Schemas

In our methods, elements in the input interface and output webpage returned by the web source are represented by their corressponding set attributes. Referring to our running example, the sub-figures a) and e) show the input interface and output pages for a data source *A*.

Input attributes are the elements on the interface that enable users to submit their queries. In most deep web data sources, the user submits queries to the data source through entering strings into the text input box on the input interface of the data source. An example of text input box is shown in Figure 2 a). Thus, in our algorithm, we consider the strings that can be input on text boxes as the input attributes. The text box usually has a label attached to it, which reveals the semantics of the input attribute and the corressponding instances. Thus, an input attribute is modeled by its label and instance set: $A = \{L, I\}$, where L refers to label and I refers to the instance set.

As with the input attributes, output attributes also have label and an instance set attached to them. Specially, in many data sources, output attributes related to each other are often presented in a table or a separate block. For example, in Figure 2, sub-figure e), Allelic, Minor Allelic Frequency as well as Heterozygosity, which are attributes related to SNP's frequency data, are presented in a separate block. In this case, to capture such relationship between the attributes, we consider a hierarchical representation. We consider the output to be a tree of elements, where the leaf node of the tree is the attribute in the schema, and the internal element corresponds to a group of attributes in the schema. The group, in turn, corresponds to a table or a separate block in the output webpage. An internal element is further associated with a label, which is the text surrounding the group or the table. Attributes with the same parent are siblings.

With this hierarchical representation, a leaf node attribute A is modeled as follows:

$$A = \{L, I, P, S\}$$

Here L is the label for the attribute, I is the instance set of the attribute, P is the label of the parent element for the attribute, and S is the set of sibling attributes.

2.2   Similarity Function

This subsection summarizes the schema matching function used in our work. This is closely based on the pervious work on schema matching [22,12]. We consider two attributes, $A_1 = \{L_1, I_1, P_1, S_1\}$ and $A_2 = \{L_2, I_2, P_2, S_2\}$. The similarity function is a weighted sum of the similarities of their components. Thus,

$$\begin{aligned}
Simi(A_1, A_2) = {} & \omega_t \times TypeSimi(A_1, A_2) + \\
& \omega_v \times ValueSimi(I_1, I_2) + \\
& \omega_d \times DomainSimi(I_1, I_2) + \\
& \omega_l \times LabelSimi(L_1, L_2) + \\
& \omega_p \times ParentSimi(P_1, P_2) + \\
& \omega_s \times SiblingSimi(S_1, S_2)
\end{aligned} \tag{1}$$

As we can see, the above expression uses a function based on the attributes themselves, two functions that are based on the instance sets, and one function each for the parents and siblings. There is a weight associated with each of these, which can be varied. We now elaborate on these six functions.

Similarity of Type: The similarity of type as well as the similarity of value and domain, which will be described later, are instance-level properties. The attributes can be divided into two types: numeric and string , based on there instances. If the two attributes have same type, their type similarity is 1, otherwise, their type similarity is 0.

Similarity of Value: Value match between the instances is an important aspect of similarity computation. We use a Best-Match algorithm [22] to compute the *Similarity of Value* score between two instance sets. In this algorithm, the similarity function between pairs of instance string from the two instance sets is the Cosine function. We repeatedly choose the pair of the instances with the largest similarity among all pairs and delete the corresponding pair from instance sets. The process is repeated until one of the instance sets is empty. Finally, the similarity between the two instance sets is the average of the largest similarity chosen at each step.

Similarity of Domain: The similarity of domain is only computed for numeric attributes to evaluate the overlap in the ranges of the available numeric instances [22].

Similarity of Label: The similarity of labels is computed using the *linguistic similarity*, which simply computes the similarity of two strings of words. In our implementation, a well-known information retrieval technique by Salton [16] is used. We use the vector space model to model each string and use the Cosine function to compute the similarity.

Similarity of Parents: The similarity of parents between two attributes is based on the similarity of the parents' labels P_1 and P_2. We use the method we had described above to find similarity between the labels.

Similarity of Siblings: The similarity between siblings is also computed by the Best-Match algorithm we had mentioned above. The similarity function between two sibling attributes is described in formula 1 by setting ω_s to 0.

In the implementation of the above methods, an important issue is *string normalization*. In biological databases we worked with, labels often contain concatenated words. Instances of some attributes are particularly represented with a fixed prefix and a number. For example, the SNP IDs have **rs** followed by a number. Thus, the labels and instances need to be normalized before they are used to compute the string similarity. Delimiters such as "_" ".", as well as changes in *type of character* (numeric or alphabet) is used to break the string into different words. For example, **rs7412** is broken into **rs** and **7412**.

3 Main Technical Approach

Given multiple data sources with interfaces, the goal of our algorithm is to find the semantic matching between input attributes and output attributes among multiple data sources. Thus, our approach has two main components. The first component aims at finding valid instances for input attributes. Having such instances allows us to obtain instances of output attributes, and thus find output schemas. Once we have instances of input and output attributes, the second component identifies the semantic correspondence between the set of all input and output attributes across different data sources. Referring back to our example, we had shown the given three query interfaces (Figure 2, sub-figures a) c), and d)) for three deep web data sources A, B and C.

3.1 Discovering Instances for Input Attributes

For a deep web data source, sample outputs can only be obtained when we can query the input interface by finding valid instances of input attributes. Furthermore, instances of input attributes and output attributes can efficiently suggest the semantic matching between them. We combine two different ideas for finding such valid instances. First, we utilize information that the web source's interface provides. Second, we also try to learn instances for input attributes from the output webpages obtained from data sources.

Instances from Input Interface. We have developed a new approach for automatically finding instances for input attributes using the information that is typically available from web pages related with the input interface provided by the data source. The key observation is that the webpage of input interface and webpages linked by the input interface always contain informative examples that help user to learn how to query the data source. Thus, these webpages, which are called help webpages in our algorithm, provide instances for input attributes. For example, on the interface shown in Figure 2 a), the data source A provides a link attached with *search hints* located after the text input box labeled with *Enter the dbSNP ID*. The help webpage linked by *search hints* is shown in Figure 2 b). In the web page, the data source A provides the example `rs4253207` which can be used for querying the text input box and the corresponding output webpage is shown in Figure 2 c). In this case, we can use some technology to extract the potential instances like `rs4253207` from the help webpages for the input attribute T labeled with L, which is the text surrounding the text input box of T.

The algorithm for automatically finding instances from the input interface is comprised of the following steps:

Identifying Potential Help Webpages: Help webpages can be retrieved by useful links on the interfaces provided by the data sources. Thus, among all the links provided by the interface, it is important to identify the useful ones that might provide instances for users. We observed that the potentially useful links for help webpages are always attached with keywords like *help*, *search hints*, *sample*, *about*, *how*, *what*, *tutorial*, and *map*. Thus, if a link on the interface is attached with text containing any of the keywords mentioned above, it is considered as a potentially useful link, and the corresponding webpage is collected as a help webpage. In addition, the webpage of interface itself may also contain instances for the input attribute, so it is also considered as a help webpage.

Locating Potential Instances: Given a set of help webpages, the location of the potential instances should be identified. On the help webpages, the instances provided by the data sources are always surrounded by the keywords like *such as*, *:*, *()*, *e.g.*, *for example*, *for instance*, *like*, as well as the label L for the input attribute T. These keywords, denoted as set K_s, positively reveal the location of potential instances on help webpages. For example, in Figure 2 b), the instance *rs4253207* follows *:* and *dbSNP ID*, which is the label of the input attribute, and is also enclosed into parenthesis. Thus, on the help webpages, sentences that contain any keyword included in the keywords set K_s are extracted. These sentences are considered as locations that might contain potential instances.

It is helpful to preprocess the label L before using it to locate potential instances. In our algorithm, *noisy* words are removed from the label. Examples of such words are: *please*, *enter*, *this*, *that*, *a*, *an*, *it*, *search*, etc. We use a list containing such kind of words and the words contained in this list are removed from the label string.

Discovering Potential Instances: We need to further identify the potential instances from the extracted sentences in the last step. In domain-specific deep data sources, the instances used to query the interface are often domain-specific terms, which are less likely to be used in other domains. For example, *rs4253207* is less frequently used in general English sources compared with other common words such as *enter*, *number*, *search*, etc. Thus, we can use this property to distinguish the potential input instances from the common words surrounding them.

The frequency of the terms appearing on some general English sources is utilized to identify the potential instances. In our implementation, a large collection of documents is crawled from the web within the six domains of economics, science, politics, arts, sports, and history. An inverted index is constructed for each term in the collection and the corresponding *document frequency df* is computed. The document frequency *df* of a term implies the number of documents in the collection that contain the word.

For each term in sentences obtained in the last step, if its document frequency *df* is less than a threshold λ, it will be considered as a potential instance for the text input box T. In our algorithm, we set the threshold λ to 20.

Validating Potential Instances: To determine whether the extracted instances are true instances or not, the potential instances are validated through the interface. Extracted instances are submitted to the text input box T on the input interface and the output webpages returned by the data source are compared with a *dummy webpage*. This dummy webpage is obtained by submitting a predefined nonsense query to the interface. If the output webpage is similar to the dummy webpage, the corresponding instance is considered false. Otherwise, the corresponding instance is considered to be a true instance for the text input box T.

Obtaining Instances from Output Webpages. It is possible that there are no instances provided by the interfaces of data sources , or the quantity of instances found from the interface is not large enough to support reliable semantic matching. In this case, more instances are required for the input attributes. Besides help webpages of the interface, another type of informative source that provides instances for an input attribute is the output webpages from other data sources in the same domain. A dynamic algorithm is utilized to agglomeratively find instances for input attributes from multiple data sources. The instances of output attributes might be able to query input attributes of other data sources if they are similar to each other, resulting in more output webpages and instances of output attribute that can further provide instances for input attributes.The similarity between output attributes and input attributes here is utilized to suggest the potential instance provision, rather than identifying the semantically matching attributes, which will be solved in the next section.

The example in Figure 2 shows the effectiveness of the algorithm. In Figure 2 c) and Figure 2 d), there is no instance for input attributes that can be found directly from the input interface. However, in the output webpage shown in Figure 2 e), which is acquired by using the instance *rs4253207* to query the interface of the data source A, an output attribute with the label of *Gene* is obtained. The instances obtained here may provide instances for the input attribute with the label of *Gene name*, which is seen on the input interface of the data source

B. This is because these two labels are very similar to each other, and further, our validation method can always prune out instances that are not correct. Similarly, the output attribute labeled with *Gene ID* in the data source B can provide instances for querying the data source C, which is shown in Figure 2 d).

In general, we can use instances from other attributes which are likely to be semantically similar to the given input attribute. The method is shown as Algorithm 3.1. The input of the algorithm is a set of input attributes IA for which we need to find the instances, the current attribute set CA , including output attributes which already have instance sets, and a threshold λ_q, which is used to control the size of input attributes' instance set. In the algorithm, input attributes $ia \in IA$ are repeatedly probed by borrowing instances from the current attribute set CA, until the size of ia's instance set, I_{ia}, is larger than λ_q. Probing input attributes means using instances of the attributes in the CA set to query the corresponding interface. In order to borrow instances for the input attribute ia, the algorithm first computes the similarity values between the attribute $ca_i \in CA$ and the input attribute ia. Then, the algorithm successively borrows instances for ia from the instance set of each ca_i in set CA, according to the decreasing order of the similarity values computed in the S set. Enriched with new attributes and instances, the CA set is able to provide more instances for input attributes in set IA. If any attribute in the IA set has an instance set larger than λ_q, the attribute is removed from the IA set since we have found enough instances for it. The boolean variable, $IsProbed$, is used to control the outer-most loop, making sure the procedure of checking new discovered instances.

Algorithm 3.1: ProbeInterface(IA, CA, λ_q)

/*IA is a set of input attributes need to be probed, CA is
the current attribute set which has been probed, λ_q is threshold*/
Set a boolean variable $IsProbed = true$
  while ($IA \neq null$ **and** $IsProbed = true$)
    Set $IsProbed = false$
    foreach $ia \in IA$
      Compute the similarity set S
      Sort the elements of S in descending order
      while $S \neq null$ **and** $|I_{ia}| < \lambda_q$
       Choose attribute ca_t which has the largest similarity $s_t \in S$
       Query ia using instances of ca_t until $|I_{ia}| > \lambda_q$
       if there are successful queries **and** $IsProbed = false$
        $IsProbed = true$;
       if $|I_{ia}| > \lambda_q$
        Remove ia from IA
       Update CA using the output webpage of ia
       Remove s_t from S

The instance set for input attribute on the unprobed interface is *NULL*, while instances are continuously added into the instance set by borrowing from other attributes. The similarity between ia and $ca_i \in CA$ is computed using the similarity function we had described earlier.

We have also observed that semantically matched attributes have overlap in their instance sets. Thus, if we borrow instances from attributes belonging to other data sources, we tend to extract the same biological objects from these data sources. For example, by borrowing instances, our algorithm tends to extract the biological object with Gene Name of *ERCC6* from data sources A, B and C, though their input attributes are different. Accordingly, from the output webpages shown in Figure 2, sub-figures e), f), and g), we can find three attributes, *Chromosome* from the data source A, *Location* from the data source B and *Gene Map Locus* from the data source C, whose semantic correspondence is greatly revealed by their overlapping instance sets. Thus, our algorithm is able to find such a match because we borrowed an output instance to query another data source, and thus, we are referring to the same biological object when querying the different data sources.

3.2 Schema Matching

Output schemas can be obtained by querying through the interface, and by analyzing the output webpages.

The algorithm for schema matching is based on a clustering process, which aims at grouping attributes that have semantic correspondence. The input of the algorithm is the set of attributes gathered from data sources, including input attributes and output attributes. The algorithm starts by constructing a cluster set, where each cluster in the set is composed of a single attribute from the attribute set. Then, it repeatedly merges the pair of clusters with the largest similarity into one cluster until there is no pair of clusters having similarity larger than a threshold ρ.

For two clusters C_i and C_j, each of which is a set of attributes: $C_i = \{A_{i1}, A_{i2}, ... A_{in}\}$ and $C_j = \{A_{j1}, A_{j2}, ... A_{jm}\}$. The similarity between these two clusters is the average of pairwise similarity between all pairs of attributes from C_i and C_j:

$$ClSimi(C_i, C_j) = \frac{\sum(Simi(A_{it}, A_{jk}))}{n * m} \qquad (2)$$

Where $A_{it} \in C_i$, $A_{jk} \in C_j$, and n, m are the sizes of cluster C_i and C_j. The similarity between two attributes $Simi(A_i, A_j)$ is calculated by similarity function we had introduced earlier.

The output of the schema matching algorithm is a cluster set. Each cluster in the set is composed of attributes with semantic correspondence. Thus, if an input attribute and an output attribute from different data sources belong to the same cluster, there is an input-output relation between these two data sources. Besides the input-output relation, our algorithm could also identify the input-input schema matches, and output-output schema matches. Further, if several attributes from the same data source and an attribute from another data source are in the same cluster, then there would be $1 : m$ mappings among these attributes, e.g. *Gene Ontology* may be map to *Gene Function*, *Gene Process*, and *Gene Component*. An attribute like *Gene Ontology* is referred to as a *composite attribute* in our work, whereas all other attributes are considered *simple attributes*.

4   Evaluation Study

Our techniques for automatically integrating deep web data sources were evaluated using a case study that focused on the SNP related biological data sources. In the effort to explain the genetic contribution to complex diseases such as cancer and heart disease, single nucleotide polymorphisms (SNPs), that designate sites in the genome that has two or more nucleotide variants segregating in a population, seem particularly promising because they are usually biallelic and thus easily assayed. Because already over fourteen million human SNPs have been reported in the public database (dbSNP, build 129), it is desirable to develop methods of sifting through this information to find likely candidates for disease association. One important prerequisite for searching SNP databases for likely candidates for disease association is to be able to link information on SNP allelic frequencies with information on the genomic location of a SNP and, if the SNP is located within a gene (and over 6 million of them are[1]), the biological roles of the protein encoded by that gene. For example, allelic variants that contribute to disease are expected to have an effect on the phenotype, either by causing a change in the amino acid sequence of a protein or by effecting regulation of gene expression and/or intron splicing. Likewise, plausible candidate SNPs for association with a given disease are likely to be found in genes whose protein products are known to contribute to the biological processes and pathways involved in the disease. Information on human SNPs is also useful for studying questions relating to human evolutionary history and the role of population genetic processes such as natural selection in shaping the human genome. Thus, providing effective search mechanisms for SNP-related information requires integration of a large number of independent biological data sources.

Table 1. Instances Discovered from Input Interfaces

Data Source	Interface	# from Interface
SNP500 Cancer	SNPID	1
	GeneName	2
dbSNP	rsID	5
	ssID	8
Gene	GeneName	4
SIFT	SNPID	10

Particularly, in this study, we integrated 11 data sources with 24 query interfaces which include SNP500Cancer, dbSNP, SeattleSNP, HumanProtein, and others. The deep web data sources provide data about SNP, Gene, Protein and related information. All the interfaces are text-query based interfaces and each interface contains one text input box, which implies that there is one input attribute on each query interface. For all experiments we conducted, the weight coefficients for the component similarities are set as follows: $\omega_t = 0.05, \omega_v = 0.22,$ $\omega_d = 0.33, \omega_l = 0.2, \omega_p = 0.1$ and $\omega_S = 0.1$. This reflects the observation that the instance-level property, including similarity of type, value and domain, is

---

[1] http://www.ncbi.nlm.nih.gov/SNP/snp_summary.cgi

very important to identify the semantics of attributes. For attributes where one or more of parent, siblings, and instances are null, the weights are adjusted to maintain the same ratio between weights of non-null properties. We conduct our experiment on all the biological data sources by setting $\rho = 0.2$, which is the stopping threshold for clustering process, and setting $\lambda_q = 30$ which is the threshold for the number of querying instances. The threshold λ_q is chosen so that the performance of the system is stable, which means that increasing λ_q won't largely increase the system's performance.

4.1   Effectiveness of the Approach

One of the important aspects of our work is the method for discovering input instances directly from the input interfaces. Table 1 summarizes the number of instances we are able to discover automatically. From the table, we can see that among the 24 input attributes, there are 6 input attributes for which we are able to obtain instances directly from interface. For all other input attributes, instances are learnt from output webpages dynamically, as we had described in Section 3.1. Thus, using the help webpages associated with these 6 input attributes, our method is able to successively find instances for all input and output attributes in the system, and enable schema matching.

As an example of the functionality of our method, Table 2 shows the input-output relations among 5 interfaces identified by our algorithm. In the table, each row corresponds to an input-output relation. The first column shows the data source that contains the input attribute, the second column is the data source that provides the output attribute and the third column is the matching attribute. The fourth column shows whether our system identifies the method correctly. As we can see, matches with a number of different attributes are identified within these 5 data sources. All but one of the matches shown here were identified correctly our system.

Table 3 shows the total count of semantic matching identified across all 24 query interfaces. We separate input-input schema matching, input-output schema matching, and output-output schema matching.

Table 2. Input-Output Relations Identified between 5 Data Sources

Input Source	Output Source	Attribute	Correct
SNP500 Cancer	Seattle	SNPID	Y
Gene	Seattle	GeneID	Y
Gene	HumanProtein	GeneID	Y
Gene	HGNC	GeneID	Y
Seattle	SNP500 Cancer	GeneName	Y
Seattle	HumanProtein	GeneName	Y
Seattle	HGNC	GeneName	Y
HumanProtein	Gene	OMIM	N
HumanProtein	HGNC	OMIM	Y
HGNC	Gene	GeneBank ID	Y

Table 3. Schema Matches Across Data Sources

Type	Discovered	Correct	Missed
In-In	21	19	0
In-Out	141	128	6
Out-Out	385	297	77
All	547	444	83

Table 4. Performance: Accuracy of Schema Matching

Type	Precision(%)	Recall(%)	F(%)
In-In	90.4	100	94.9
In-Out	90.7	95.5	93.0
Out-Out	77.1	79.4	78.8
All	81.2	84.2	82.6

4.2   Quantitative Analysis

We use three metrics to quantify our approach: *precision, recall,* and *F-measure.* Precision is the percentage of the correct mappings over all mappings identified by our method. Recall is the percentage of the correct mappings identified by our method over all mappings in the data set given by the domain expert. F-measure incorporates both precision and recall with the following expression

$$F = 2PR/(P + R)$$

where, F, P, and R denote the F-measure, Precision, and Recall, respectively.

Table 4 shows the performance of our system on overall schema matching as well as on different types of schema matching we perform. The result on overall schema matching shows that our automatic matching method is effective. With some assistance from domain experts, our techniques can help achieve accurate integration across data sources. From the table we can also see that our system has good performance on input-output attribute mapping, which implies that our system can efficiently discover the input-output relations between data sources. However, our system does not perform so well on output-output attribute mapping. This is because the output attributes are often more complex than the input attributes, which make it more challenging to identify the output-output mappings.

We also conducted a number of experiments on overall schema matching to further understand the cases in which our techniques are not effective. First, we focused on understanding the relative performance of our techniques on two types of attributes, i.e, the *simple attributes* and *composite attributes*. As we explained earlier, Composite attribute refers to an attribute that may be mapped to more than one attributes in another data source, while a simple attribute is only mapped to one attribute in other data sources. We divide our data sources into two sets. The first set is composed of the data sources which only have simple attributes in their schemas, while the second set is composed of the data sources which contains composite attributes in their schemas.

The performance of our system on these two sets is shown in Table 5. From the results, we can see that our system has better performance on simple attributes. In other words, it is important to improve our method on composite attributes to further achieve better integration.

We also evaluated the performance of our method based on different attribute types, namely *string attributes* and *numeric attributes*. For an attribute, if all the instances are numeric, it is called a numeric attribute, otherwise, it is called a string attribute. The performance of our system on numeric and string attributes is shown in Table 6.

Table 5. Performance on Simple and Composite Attributes

	Precision(%)	Recall(%)	F(%)
Simple	83.0	88.0	85.4
Composite	82.2	71.4	76.4

Table 6. Performance: String and Numeric Attributes

	Precision(%)	Recall(%)	F(%)
String	86.9	84.2	85.52
Numeric	95.0	97.4	95.7

Compared with the performance on the entire data set, our system performs better on each individual type of attribute. This result illustrates that it is more difficult to perform schema matching on mixed types of attributes than single type of attributes. The table also shows that the system has better performance on numeric attribute than string attribute. This is because most of the composite attributes are string attributes, which deteriorates the performance of our approach.

5   Related Work

Schema matching is an important problem in data integration, and has been very well studied [8,9,12,10,22,17,20,19,21]. Early systems for scheme matching[8,12] focused on traditional databases. These systems mainly compare the pairwise similarity between attributes, based on their labels and instances, the structures of the schema, as well as any constraints. The LSD system [8] uses machine-learning methods to match a new schema with a predefined global schema. Cupid [12] is a generic schema matching system that discovers matches in schema elements based on their names, data types, constraints and schema structure.

In recent years, many efforts have focused on schema matching on input interfaces of deep web data sources[9,10,22]. The main goal of these system is to provide a unified query interface for the user. He *et al.* [9] gave the hypothesis that there is an underlying generative global schema model for the input interfaces across different web data sources in each domain. He *et al.* [10] introduced a system called WISE-Integrator, which performs automatic integration of web interfaces. Wu *et al.*[22] propose an interactive clustering-based approach for integrate query interfaces of data sources. The algorithm identifies mapping attributes by clustering them into a group based on their names and values. The difference between our work and these systems is that they do not consider output attributes. Sarma *et al.*[17] have proposed a Pay-As-You-Go system to perform schema matching between the data sources and a mediated schema. A probabilistic mediated schema is created in this system and probabilistic schema mappings between the sources and the mediated schema is identified automatically.

More closely related to our work, Wang *et al.* [20] have developed a schema matching system that works on both the input and output schemas of deep web data sources. The system maps attributes based on comparing their instances with the attributes in a predefined global schema. The instances are obtained by *probing* the interface of deep web data sources. The probing method they use is based on assuming availability of a domain-specific global schema, and a set of pre-defined data objects under this schema. Our approach is clearly distinct, as we find query instances from the help pages, and further, use instances from other similar attributes obtained dynamically from output web pages. WebIQ[21] is a system that discovers instances for input attributes from the surface Web and the Deep web. WebIQ extends question answering techniques commonly used in the AI community and discovers instances from a surface search engine. Our work is based on use of existing help links, which are not used in their work. WebIQ also borrows instances from other attributes, though the exact algorithm used in our work is very different.

With rapid increase in the number of biological data sources, several recent efforts have focused on biological schema matching [11,19]. Wang et al. [19] compare the attributes in different schemas based on their names and the schema structure with a new emphasis on using the context of the attribute provided in the schema structure.

The main novel aspect of our work is that we perform schema matching on *inter-dependent* deep web data sources, and perform an application study focusing on biological deep web data sources. None of the previous work on schema matching on deep web data sources considered either of these aspects: i.e. interdependent data sources, or biological data sources. Our algorithm automatically discovers the instances for input attributes, generates the global schema, and identifies matching attributes.

Somewhat related to our work, there are systems designed for probing and extracting data from deep web data sources [7,4,13]. The systems proposed by Callan et al [7] and Barbosa et al [4] focus on extracting documents from text databases through keyword-based query interfaces. Similar as our strategy, their systems also use the outputs from previous probes as new input instances for the new probe. However, the input schema of text databases has fewer constraints, i.e., all English words are valid input instances. As a result, unlike our method, they do not need to match the output with the input.

6   Conclusion

In this paper, we have applied a series of data mining techniques for schema matching. Particularly, in schema matching, we have considered input-output attribute matches, in addition to input-input and output-output matches. We show that the instances for querying deep web data sources can be discovered from the information provided by the *query interfaces themselves*, as well as the obtained output pages of related data sources by query probing using *dynamically* identified input instances. Our approach has been effective on a number of biological data sources.

References

1. Brookes, A.J.: The essence of snps. Gene. 234, 177–186 (1999)
2. Ashish, N., Knoblock, C.A.: Semi-automatic wrapper generation for internet information sources. In: Proceedings of the Second IFCIS International Conference on Cooperative Information Systems. IEEE Computer Society, Los Alamitos (1997)
3. Babu, P.A., Boddepalli, R., Lakshmi, V.V., Rao, G.N.: Dod: Database of databases–updated molecular biology databases. Silico Biol. 5 (2005)
4. Barbosa, L., Freire, J.: Siphoning hidden-web data through keyword-based interfaces. In: Proceedings of SDDB (2004)

5. Bergman, M.K.: The deep web: Surfacing hidden value. Journal of Electronic Publishing 7(1) (August 2001)
6. Buneman, P., Davidson, S.B., Hart, K., Overton, C., Wong, L.: A data transformation system for biological data sources. In: Proceedings of the Twenty-first International Conference on Very Large Databases (1995)
7. Callan, J.: Query-based sampling of text databases. ACM Transactions on Information Systems 19, 97–130 (2001)
8. Doan, A., Domingos, P., Halevy, A.: Reconciling schemas of disparate data sources: A machine-learning approach. In: SIGMOD Conference, pp. 509–520 (2001)
9. He, B.: Statistical schema matching across web query interfaces. In: SIGMOD Conference, pp. 217–228 (2003)
10. He, H., Meng, W., Yu, C., Wu, Z.: Wise-integrator: a system for extracting and integrating complex web search interfaces of the deep web. In: VLDB 2005: Proceedings of the 31st international conference on Very large data bases, pp. 1314–1317. VLDB Endowment (2005)
11. Hern, T., Kambhampati, S.: Integration of biological sources: Current systems and challenges ahead. Sigmod Record 33, 51–60 (2004)
12. Madhavan, J., Bernstein, P.A., Rahm, E.: Generic schema matching with cupid. The VLDB Journal, 49–58 (2001)
13. Madhavan, J., Ko, D., Kot, L., Ganapathy, V., Rasmussen, A., Halevy, A.: Google's Deep Web Crawl. VLDB Endowment 1, 1241–1252 (2008)
14. Nie, Z., Wen, J.-R., Ma, W.-Y.: Object-level vertical search. In: Proceedings of the 3rd Biennial Conference on Innovative Data Systems Research, pp. 235–246 (2007)
15. Rahm, E., Bernstein, P.A.: A survey of approaches to automatic schema matching. VLDB Journal 10(2001) (2001)
16. Salton, G., Mcgill, M.J.: Introduction to Modern Information Retrieval. McGraw-Hill, Inc., New York (1986)
17. Sarma, A.D., Dong, X., Halevy, A.: Bootstrapping pay-as-you-go data integration systems. In: SIGMOD 2008: Proceedings of the 2008 ACM SIGMOD international conference on Management of data, pp. 861–874. ACM, New York (2008)
18. Wang, F., Agrawal, G., Jin, R.: Query planning for searching inter-dependent deep-web databases. In: Ludäscher, B., Mamoulis, N. (eds.) SSDBM 2008. LNCS, vol. 5069, pp. 24–41. Springer, Heidelberg (2008)
19. Wang, G., Goguen, J., Nam, Y.k., Lin, K.: Interactive schema matching with semantic functions. In: Yu, J.X., Lin, X., Lu, H., Zhang, Y. (eds.) APWeb 2004. LNCS, vol. 3007, pp. 654–664. Springer, Heidelberg (2004)
20. Wang, J., Wen, J.-R., Lochovsky, F., Ma, W.-Y.: Instance-based schema matching for web databases by domain-specific query probing. In: VLDB 2004: Proceedings of the Thirtieth international conference on Very large data bases, pp. 408–419. VLDB Endowment (2004)
21. Wu, W., Doan, A., Yu, C.: Webiq: Learning from the web to match deep-web query interfaces. In: International Conference on Data Engineering, p. 44 (2006)
22. Wu, W., Yu, C., Doan, A., Meng, W.: An interactive clustering-based approach to integrating source query interfaces on the deep web. In: SIGMOD 2004: Proceedings of the 2004 ACM SIGMOD international conference on Management of data, pp. 95–106. ACM Press, New York (2004)

Integrative Information Management for Systems Biology

Neil Swainston, Daniel Jameson, Peter Li, Irena Spasic, Pedro Mendes,
and Norman W. Paton

School of Computer Science, University of Manchester,
Oxford Road, Manchester M13 9PL, UK
{neil.swainston,daniel.jameson,peter.li,i.spasic,
pedro.mendes,npaton}@manchester.ac.uk

Abstract. Systems biology develops mathematical models of biological systems that seek to explain, or better still predict, how the system behaves. In bottom-up systems biology, systematic quantitative experimentation is carried out to obtain the data required to parameterize models, which can then be analyzed and simulated. This paper describes an approach to integrated information management that supports bottom-up systems biology, with a view to automating, or at least minimizing the manual effort required during, creation of quantitative models from qualitative models and experimental data. Automating the process makes model construction more systematic, supports good practice at all stages in the pipeline, and allows timely integration of high throughput experimental results into models.

Keywords: computational systems biology, workflow.

1 Introduction and Motivation

Systems biology involves the development and study of mathematical models of biological systems. Existing databases of pathways, combined with the emergence of consensus models of specific organisms [1], provide broad access to qualitative models of biological systems. In bottom-up systems biology, these qualitative models can be used as a starting point for the creation of quantitative models that support a range of different forms of simulation and analysis [2]. As such, bottom-up systems biology projects: (i) identify the pathway or portion of a network that is to be modeled; (ii) associate the model with functions and parameter values that represent its dynamic behavior, either from databases [3] or experimentation; and (iii) analyze and/or simulate the resulting model to understand its properties.

In common practice, model construction is a manual process, in which a modeler manually associates a qualitative model with dynamics, and experiments with the resulting model using software tools such as Copasi [4]. Such an approach can give rise to good quality models, but certainly can be seen more as a cottage industry than

P. Lambrix and G. Kemp (Eds.): DILS 2010, LNBI 6254, pp. 164–178, 2010.
© Springer-Verlag Berlin Heidelberg 2010

as a highly scaleable production process. The widespread use of high throughput experimental methods means that manual modeling can easily become the rate limiting step, and the diversity of available data sets means that modelers operate in a complex information space in which the provenance of a model can be difficult to decipher. As such, there seems to be value in exploring the extent to which the association of models with experimental data – in essence the transition from machines to models to simulations – can be automated.

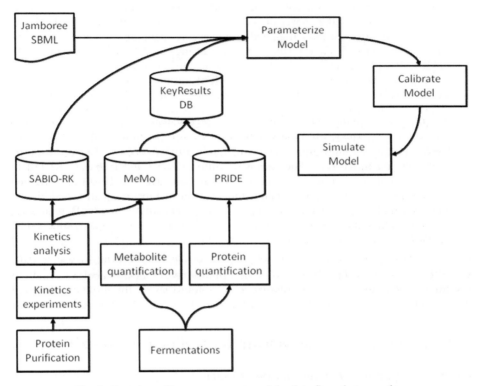

Fig. 1. Overview of key components and the data flows between them

This paper presents an approach to the automation of experimental data capture and integration with models, in which quantitative proteomics, metabolomics and reaction kinetics experimental data is used to parameterize workflows from a metabolic reconstruction, for simulation and analysis using Copasi [4]. In this approach, which is illustrated in Figure 1, the following steps take place:

1. Experimental data is captured directly from instruments, and subject to primary analyses (for example, in proteomics to obtain protein concentrations from mass spectrometry results [5]).

2. Experimental data from instruments, along with the results of the primary analyses, are archived in experimental data repositories, specifically MeMo [6], PRIDE [7] and SABIO-RK [3] that provide a comprehensive record of the experimental processes followed and the results obtained. As such, the experimental data repositories contain the levels of detail that would allow primary analyses such as protein quantifications to be rerun, and the experimental design to be validated.

3. The information required for modeling is extracted from the experimental data resources and stored in a Key Results Database (KRDB), which essentially associates sample information with experimental factors and measured results. Thus the KRDB contains the subset of the data from the experimental repositories at (2) that is required for modeling, and provides consistent representation of quantitative experimental data results for use during model parameterization. As SABIO-RK [3] was essentially already designed to support modeling tasks, we do not replicate reaction kinetics data in the KRDB.

4. A Taverna [8] workflow obtains qualitative model information, represented using SBML [9], parameterizes this model with results in the KRDB, and conveys the resulting quantitative model to the Copasi Web Service [10] for calibration and simulation. An SBML document is built up incrementally through the workflow cycle. Initially an unparameterized, stoichiometric network is extracted from the metabolic reconstruction. The parameterization workflow queries the KRDB to extract initial concentrations of metabolites and enzymes. SABIO-RK is queried allowing each reaction to be expanded in terms of its kinetic equation and kinetic parameters. These parameters are then tuned by the calibration workflow producing a model that may be submitted to the simulation workflow, generating results in SBRML format [11].

The remainder of the paper is structured as follows. Section 2 drills down on the individual components within the lifecycle, describing the key design decisions and the resulting capabilities. Section 3 presents some conclusions on the results to date and their significance for systems biology in practice.

2 Components and Characteristics

2.1 From Equipment to Experimental Results

Three experimental techniques are required to provide data for the parameterization of kinetic metabolic models: quantitative proteomics and metabolomics, and enzyme kinetic assays. In each case there is a requirement to:

(i)   perform analyses on the raw experimental data to derive the secondary quantitative parameters required in the model;

(ii)   store the raw experimental data along with relevant metadata and the derived parameters, thus providing the facility to trace back and reanalyze raw data should this be required from model simulation results. Where possible, existing data standards and tools are reused, both to reduce wheel-reinvention and development time, and also to provide the facility of sharing experimental data in formats with which the bioinformatics community is familiar.

Quantitative proteomics studies are performed using tandem mass spectrometry, utilizing the QconCAT approach in which isotopically-labelled peptides of known concentration are spiked into a sample, and peak area / intensity comparisons are used to infer peptide and therefore protein concentration in the sample [5].

In order to facilitate the analysis task, a wizard has been produced [12] that automates the steps of:

(i)   performing a database search against the Mascot search engine [13] to identify both isotopically-labelled and native peptides;

(ii)   determining which peptides can be reliably quantified, based on Mascot significance scores and peptide ion retention times;

(iii)   performing the quantification, using an algorithm developed for the SILACAnalyser tool [14];

(iv)   formatting both identification and quantification results according to the PRIDE XML data format; and

(v)   uploading both the derived results and experimental data to a native XML database.

Following submission to the XML database, data can be extracted and queried via a web and web service interface. It is standard practice in quantitative proteomics for multiple experimental replicates to be performed, and thus quantifications at the protein level are calculated using contributions from individual peptides from each replicate. These can then be queried to parameterize SBML models or extracted to populate the KRDB (see Figure 2).

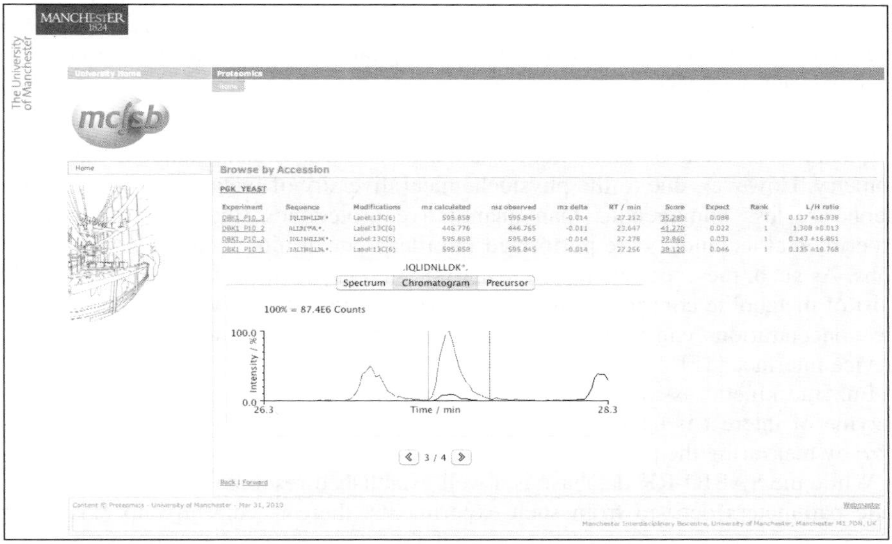

Fig. 2. Web interface displaying raw experimental proteomics data. Absolute protein concentrations can be exported from this data and stored in the Key Results Database.

168   N. Swainston et al.

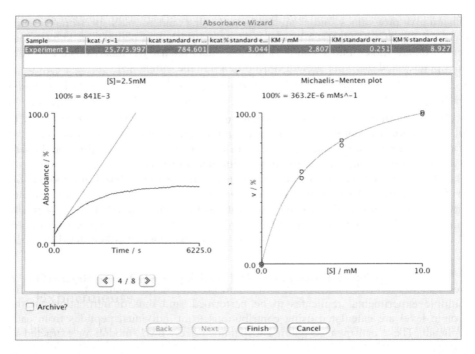

Fig. 3. The KineticsWizard result panel. The left panel displays the raw absorbance data as acquired, upon which the reaction initial rate has been fit. The initial rates from all acquisitions are plotted against substrate concentration in a Michaelis-Menten [17] plot on the right hand side. From this Michaelis-Menten plot, the kinetic parameters k_{cat} and K_M are calculated and displayed in the top panel.

Quantitative metabolomics studies are also performed using tandem mass spectrometry. However, due to the physiochemical diversity of metabolites, these experiments are less homogenous than quantitative proteomics studies, and a range of experimental techniques are performed in order to determine their *in vivo* concentrations. As such, the experimental data analysis step is performed manually, generating a list of metabolite concentrations that can be input into the MeMo database. Metabolite concentrations can then be accessed through a Pierre-generated web and web service interface [15].

Enzyme kinetic assays are performed through spectrophotometry, in which each enzyme of interest is expressed and purified, and the rate of its action measured *in vitro* by measuring the production of reaction product over time.

While the SABIO-RK database is a well-established resource for the storage of kinetic parameters derived from such experiments, there is currently no existing resource for the management of the original time course raw data from which these parameters are derived. As such, a solution has been developed in which derived parameters are stored in SABIO-RK, while associated raw data is managed in an extension of the MeMo database. Both resources are linked through web and web service interfaces, allowing the user of a given parameter to view and extract the original raw data from which the parameter was derived [16].

With a view to automating experimental data capture, analysis and deposition as much as possible, several wizards have been developed. For example, the KineticsWizard is integrated with the instrument software that performs the tasks of:

(i)   calculating kinetic parameters by applying a fitting algorithm to the time course data (see Figure 3);

(ii)   capturing sufficient metadata to allow mapping of parameters to models; and

(iii)   submitting data to the MeMo and SABIO-RK databases. SABIO-RK provides the facility to export kinetic parameters in SBML format, and can act as a single unified interface both to newly measured in-house parameters and to existing third party kinetic data.

Taken as a whole, the above resources provide the facility for analyzing and managing experimental data in such a way that derived values and parameters may be readily imported into systems biology models.

2.2   From Experimental Results to Key Results

MeMo and PRIDE store experimental data in quite specific formats that preserve information about their acquisition and subsequent processing. Within our workflow, we only require the cellular concentrations of metabolites and enzymes alongside measured kinetic parameters to be able to parameterize our models. The diverse nature of the representation of data in the independently developed experimental results repositories (MeMo, PRIDE, SABIO-RK) lends itself to being consolidated into a single repository to ease interactions with the workflows that parameterize and calibrate the model.

The KRDB [18], the data model for which is illustrated in Figure 4, allows for the amount of metadata associated with a recorded result to be reduced to the minimum necessary to support model development, and thereby facilitates the storage of these results in a set format, no matter which type of experiment they were acquired from. We note that there is not really, therefore, a single *minimum information* requirement for a type of experimental data; rather different users of experimental data results have different requirements. In this context, the target users of the experimental repositories are principally experimentalists who need to understand in detail the process through which results were produced, for example to inform reanalysis. By contrast the target users of the KRDB are modelers, who rarely have the inclination (or perhaps expertise) to make full use of the details captured in the experimental data repositories.

We view a result as a particular reading, or calculated quantity (Measurement) of a particular thing (MeasuredItem, MeasuredItemType), gathered under a particular set of conditions and possibly at a particular time (in the data model these conditions are referred to as Factors). These conditions may be either static

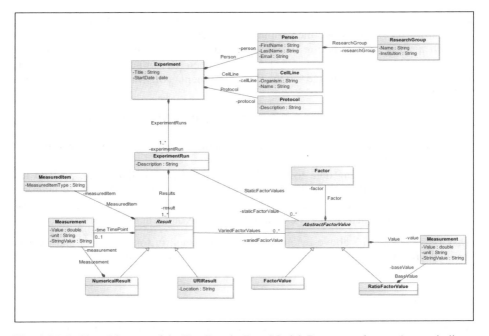

Fig. 4. UML Class Diagram of the Key Results Data Model. Boxes are classes. Arrows indicate classes that have inherited from a parent class. Lines with diamonds indicate classes that are possessed by another class – the diamond end of the line extends from the containing class. Labels indicate what classes represent in the containing class and the cardinality of the relationships. All relationships are 1:1 unless otherwise indicated (0-* is any number of instances or none at all, 1-* is at least one instance).

throughout the experiment (`StaticFactors`) or vary as the readings taken from the experiment progress (`VariedFactors`).

The data model has been converted into an XML schema, and a repository for documents conforming to this schema has been developed. Data for submission to the KRDB is formatted using spreadsheet software into tab delimited files, consisting of a list of measurements and the variable factors associated with each measurement. Once in this format a web form is used to add additional annotation to the experiment and submit the data to the database.

The web form allows basic details of the experiment to be entered, along with any specific conditions surrounding the experiment (the `StaticFactors` described earlier). Upon submission, the form and tab delimited file are checked for basic consistency and then processed by server side software into an XML document conforming to the KRDB schema. This document is then stored in eXist, an open-source, freely available XML database that provides support for web and web service interface development (http://exist-db.org/).

Data may be retrieved from the repository either manually through a web-based front end, or as in the case of our workflows, by using eXist's RESTful web service interface.

For our workflow we store the consolidated cellular concentrations of metabolites and proteins in the KRDB. These numbers are calculated by the experimentalists using the Wizards described in Section 2.1 from their replicate data and recorded in tab delimited format as described above. Metabolites are identified by ChEBI IDs [19], unambiguously linking the metabolite to a defined chemical structure, and likewise enzymes by gene or protein identifiers, such as Saccharomyces Genome Database (SGD) [20] and UniProt [21] respectively. Figure 5 shows a fragment of KRDB XML describing the concentration of the enzyme YCR012W.

```
<CellLine>
    <Name>Y23925</Name>
    <Organism name="Saccharomyces cerevisiae"/>
</CellLine>
<ExperimentRun>
    <TypeOfResult name="Protein Quantification"/>
    <Description>Proteomics quantifications</Description>
    <NumericalResult>
        <VariedFactorValues/>
        <MeasuredItem
            itemType="YeastGeneAccession">YCR012W</MeasuredItem>
        <Measurement unit="cell^-1">2818332.709</Measurement>
    </NumericalResult>
```

Fig. 5. Fragment of KRDB XML showing markup for the quantification of the enzyme YCR012W, measured in the unit copy numbers per cell.

2.3   From Results to Parameterized Models

The three sets of results produced from experiments measuring the activity of enzymes and their concentrations, as well as those of the metabolites involved in enzymatic reactions, are used for the parameterization of systems biology models. Integration of the experimental data with initially qualitative models can be achieved in a systematic manner using procedures constructed and enacted by a workflow management system such as Taverna [8]. These workflows define the flow of data between computational resources, which have been deployed as web services, enabling databases such as SABIO-RK and the KRDB to be queried by the workflow enactment engine.

Taverna workflows have been written to assemble, optimize and simulate parameterized systems biology models. These steps are characterized by successive transformations of a SBML model with quantitative data. Parameterization of a systems biology model initially requires a skeleton SBML model that describes, in a qualitative fashion, the components and their relationships with one another in a biological system. In terms of a metabolic pathway, metabolites and enzymes represent the nodes of this system, whilst the edges between these components represent biochemical reactions. Information about individual metabolic reactions originates from a web service providing access to a metabolic reconstruction. Smaller, more manageable, models comprising specific metabolic pathways are constructed by a qualitative model construction workflow based on some given criteria such as a list of enzyme names, as illustrated in Figure 6. Various metadata are retrieved for enzymes and metabolites from the consensus metabolic model web service. These metadata include references to external databases including ChEBI and UniProt) identifiers, so that metabolites and enzymes can be uniquely identified within a SBML model. Information representing the association of metabolites and enzymes for each reaction is then retrieved from the metabolic reconstruction web service. Based on this collated data, an SBML document is assembled using classes and methods from libSBML by the qualitative model construction workflow [22].

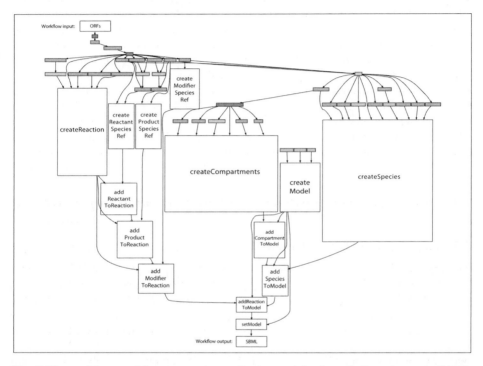

Fig. 6. The workflow used for constructing qualitative models of metabolic pathways in SBML. Calls to the consensus network web service (grey boxes) provide information about the protein, the catalysed reaction and its constituent metabolites for each enzyme from a list of open reading frame numbers. This information is used within nested workflows (white boxes) to iteratively generate components in SBML models using methods from libSBML. An SBML model is produced as the output of the workflow.

The creation of a qualitative SBML model defining how components are related to one another in metabolic reactions provides a context for the integration of proteomics, metabolomics and reaction kinetics data. The parameterization of the SBML model undertaken by this second workflow involves the mapping of quantitative experimental data onto the model which is dependent upon the external database identifiers that have been used to reference metabolites and enzymes in the qualitative SBML model, and the key results and SABIO-RK databases (Figure 7).

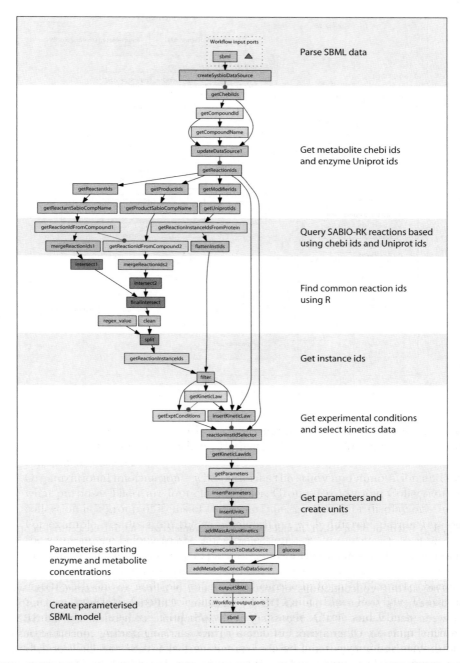

Fig. 7. Model parameterisation workflow integrating experimental data from SABIO-RK and the KRDB with a qualitative SBML model. Quantitative data from SABIO-RK and the KRDB were used to parameterise source metabolites and enzymes with their starting concentrations, and reactions with enzyme kinetics.

```
<sbml>
 <model>
   ...
   <listOfCompartments>...</listOfCompartments>
   <listOfSpecies>
    <species id="M_172" name="ATP" initialConcentration="3.5"/>
    <species id="E_670" name="YCR012W" initialConcentration="0.055110250684886"/>
     ...
   </listOfSpecies>
   <listOfReactions>
    <reaction id="R_1023" name="phosphoglycerate kinase">
     <listOfReactants>
       <speciesReference species="M_4"/>
       <speciesReference species="M_135"/>
     </listOfReactants>
     <listOfProducts>
       <speciesReference species="M_57"/>
       <speciesReference species="M_172"/>
     </listOfProducts>
     <listOfModifiers>
       <modifierSpeciesReference species="E_670"/>
     </listOfModifiers>
     <kineticLaw>
      <math xmlns="http://www.w3.org/1998/Math/MathML">
       <apply>
         <divide/>
          <apply>
            <times/>
             <ci> E_670 </ci>
             <ci> kcat </ci>
             <ci> M_57 </ci>
          </apply>
          <apply>
            <plus/>
             <ci> Km </ci>
             <ci> M_57 </ci>
          </apply>
       </apply>
      </math>
      <listOfParameters>
       <parameter id="kcat" value="343.5"/>
       <parameter id="Km" value="0.77"/>
      </listOfParameters>
     </kineticLaw>
    </reaction>
    ...
 </sbml>
```

Fig. 8. A fragment of an SBML model showing the parameterized starting concentration of the enzyme labeled as YCR012W, This concentration was calculated by the parameterisation workflow using the data shown in Figure 5

The starting concentrations of metabolites and enzymes are parameterized with measurements stored in the KRDB by matching ChEBI and UniProt identifiers between data values in this repository with appropriate components in the SBML model (Figure 8). Other sources of data for parameterising starting concentrations of metabolites and enzymes can be used, providing that ChEBI and UniProt database identifiers have been used to reference measurements. In contrast, the combination of ChEBI and UniProt identifiers that defines each metabolic reaction in the qualitative SBML model is used by the parameterisation workflow to search for relevant kinetics in the SABIO-RK database. It is often the case that this search results in multiple sets of kinetic data being found for a given reaction due to readings that have been measured for enzymes under different assay conditions. In these cases, the parameterisation workflow invites the user to select those kinetics required for the reaction based on experiment conditions. The output of the parameterisation workflow is a SBML document whose reactions have been parameterised with reaction kinetics and starting concentrations for source metabolites.

2.4   From Parameterised Models to Simulation Results

Prior to their use in predictive studies, parameterized models may be optimized in order to improve their accuracy when used in simulations. Experimental measurements of metabolite concentrations can be used to modify parameters in reaction kinetics until the output of the model produces results similar to those obtained from experimentation. Optimization of the SBML model can be performed in workflows by making use of an optimization algorithm in COPASI that has been exposed as a web service [10]. However, optimization of an SBML model is a complicated process. Firstly, there is the problem with mapping metabolomics measurements with metabolites in systems biology models. Secondly, the process of model optimization requires selection of those parameters to optimize and to what extent. A model calibration workflow has been implemented which converts experimental data into SBRML, thus allowing metabolomic measurements to be associated with components in SBML models [11]. This workflow also features the use of a pop up window wizard that invites the user to configure those parameters requiring optimization.

The calibration workflow uses the COPASIWS optimization web service that has been implemented in an asynchronous fashion due to the compute-intensive nature of the process. The workflow initiates a request for a job identifier that is then used to ensure that data is loaded and configured appropriately for each optimization process. The output of the calibration workflow is a SBML model whose original reaction parameter values have been modified against metabolomics measurements by the COPASIWS optimization web service. This optimized SBML model may now be used for simulation, which can also be performed using workflows. Such a workflow was constructed which invokes the time course simulation service available from COPASIWS, which can provide results in SBRML format for further processing (Figure 9). For example, the time course results for specific metabolites can be extracted and then used to plot how concentration varies according to time.

Whilst workflows can automate the integration of data, manual perusal of the models between each workflow stage of the process of generating optimized systems biology models is required to ensure that they make sense from a biological point of view. The set of systems biology workflows is reliant on data and metadata in all databases being consistent, otherwise anomalous models can be generated. For example, the presence of charge-balancing protons in reactions from one database but not for the same reaction in another source can lead to the inability of workflows to parameterise reactions. Problems with parameterisation of starting concentrations can also occur when the same metabolites have been referenced with different ChEBI identifiers in databases.

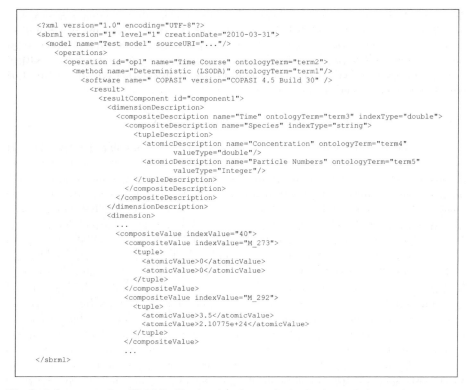

```
<?xml version="1.0" encoding="UTF-8"?>
<sbrml version="1" level="1" creationDate="2010-03-31">
  <model name="Test model" sourceURI="..."/>
    <operations>
      <operation id="op1" name="Time Course" ontologyTerm="term2">
        <method name="Deterministic (LSODA)" ontologyTerm="term1"/>
          <software name=" COPASI" version="COPASI 4.5 Build 30" />
            <result>
              <resultComponent id="component1">
                <dimensionDescription>
                  <compositeDescription name="Time" ontologyTerm="term3" indexType="double">
                    <compositeDescription name="Species" indexType="string">
                      <tupleDescription>
                        <atomicDescription name="Concentration" ontologyTerm="term4"
                            valueType="double"/>
                        <atomicDescription name="Particle Numbers" ontologyTerm="term5"
                            valueType="Integer"/>
                      </tupleDescription>
                    </compositeDescription>
                  </compositeDescription>
                </dimensionDescription>
                <dimension>
                  ...
                  <compositeValue indexValue="40">
                    <compositeValue indexValue="M_273">
                      <tuple>
                        <atomicValue>0</atomicValue>
                        <atomicValue>0</atomicValue>
                      </tuple>
                    </compositeValue>
                    <compositeValue indexValue="M_292">
                      <tuple>
                        <atomicValue>3.5</atomicValue>
                        <atomicValue>2.10775e+24</atomicValue>
                      </tuple>
                    </compositeValue>
                    ...
  </sbrml>
```

Fig. 9. A fragment of an SBRML file showing the results in the form of changes in metabolite concentrations produced by the time course simulation workflow

3 Conclusions

The construction of predictive metabolic models from experimental data is a considerable bottleneck in high throughput Systems Biology. Our information management strategy has defined a process (Figure 1) whereby data may be acquired and used in a systematic, and largely automated way, go some way towards alleviating this limiting step.

The initial hurdle for any information management strategy is not the development of repositories, but the streamlining of the process of data acquisition in such a way that those generating it do not perceive it as burdensome. Adding value during the process of data acquisition shifts the benefits from some detached individual somewhere down the line to one that is of obvious relevance to the experimentalist. Our "wizard" acquisition tools perform precisely this function; the PrideWizard, as well as capturing metadata associated with experiments, automates the quantification of peptides within minutes rather than the days it takes to perform this task by hand. The KineticsWizard also captures metadata, but automates the calculation of kinetic parameters, removing the need to perform this task by hand.

Once acquired, the data must be made available to consumers. These consumers will either be modelers or, in the case illustrated in this paper, modeling workflows.

We are aware that the consumers may not have access to the resources needed to implement or maintain the heavyweight experimental databases (MeMo, PRIDE) that we have used to archive our data. To this end, the KRDB was developed to integrate data from MeMo and PRIDE, and make it available to systems biology workflows. The use of the KRDB makes our data and workflows both useful to and implementable by others. Utilisation of such generic resources ensures that the system is applicable to a range of organisms. Although yeast is studied in this demonstration, none of the tools used are limited to this organism.

The Taverna workflows that assemble, optimize and simulate systems biology models are the consumers of the acquired data, and are the culmination of the information management workflow. By automating these processes we have tackled the perceived rate-limiting step of model construction, and provided the facility to run repeated simulations incorporating automatically selected parameter values.

The facility to simulate multiple parameter sets quickly is valuable. It allows multiple hypotheses to be tested *in silico*, which may then inform experiments that need to be conducted in order to validate the generated models, or highlight elements of metabolism that may be manipulated experimentally to further understanding.

As high-throughput experimental techniques continue to improve, the problem of how to manage the data generated will be a perpetual one. The workflow, tools and repositories presented and described here demonstrate how integrated information management can support and expedite the processes of integrative systems biology, something that can only become of increasing importance.

Availability

All workflows and accompanying documentation are available from myExperiment at http://www.myexperiment.org/packs/107. The Taverna workbench (version 2.1) can be downloaded from http://www.taverna.org.uk to run workflows, which make use of a key results database available from http://beaconw.cs.manchester.ac.uk:8780/mcisbkrdb/ and SABIORK that is accessible at http://sabio.villa-bosch.de. The COPASI web service is available from http://www.comp-sysbio.org/CopasiWS/.

Acknowledgements

The authors thank the EPSRC and BBSRC for their funding of all authors through the Manchester Centre for Integrative Systems Biology grant (BBSRC/EPSRC Grant BB/C008219/1) and thank the BBSRC for the funding of Daniel Jameson through the Dynamics and function of the NF-κB signalling system SABR grant (BB/F005938, BB/F00561X).

References

1. Herrgard, M.J., et al.: A consensus yeast metabolic network reconstruction obtained from a community approach to systems biology. Nat. Biotechnol. 26(10), 1155–1160 (2008)
2. Klipp, E., Herwig, R., Kowald, A., Wierling, C., Lehrach, H.: Systems Biology in practice. Wiley, Chichester (2005)

3. Rojas, I., Golebiewski, M., Kania, R., Krebs, O., Mir, S., Weidemann, A., Wittig, U.: Storing and annotating of kinetic data. Silico Biol. 7 (suppl. 2), S37–S44 (2007)
4. Mendes, P., Hoops, S., Sahle, S., Gauges, R., Dada, J., Kummer, U.: Computational modeling of biochemical networks using COPASI. Methods Mol. Biol. 500, 17–59 (2009)
5. Rivers, J., Simpson, D.M., Robertson, D.H., Gaskell, S.J., Beynon, R.J.: Absolute multiplexed quantitative analysis of protein expression during muscle development using QconCAT. Mol. Cell. Proteomics 6(8), 1416–1427 (2007) (Epub May 17, 2007)
6. Spasic, I., et al.: MeMo: a hybrid SQL/XML approach to metabolomic data management for functional genomics. BMC Bioinformatics 7, 281 (2006)
7. Jones, P., Côté, R.G., Martens, L., Quinn, A.F., Taylor, C.F., Derache, W., Hermjakob, H., Apweiler, R.: PRIDE: a public repository of protein and peptide identifications for the proteomics community. Nucleic Acids Res. 34, D659–D663 (2006)
8. Oinn, T., et al.: Taverna: a tool for the composition and enactment of bioinformatics workflows. Bioinformatics 20, 3045–3054 (2004)
9. Hucka, M., et al.: The systems biology markup language (SBML): a medium for representation and exchange of biochemical network models. Bioinformatics 19(4), 524–531 (2003)
10. Dada, J.O., Mendes, P.: Design and Architecture of Web Services for Simulation of Biochemical Systems. In: Paton, N.W., Missier, P., Hedeler, C. (eds.) DILS 2009. LNCS, vol. 5647, pp. 182–195. Springer, Heidelberg (2009)
11. Dada, J.O., Spasic, I., Paton, N.W., Mendes, P.: SBRML: a markup language for associating systems biology data with models. Bioinformatics (Feburary 21, 2010) (Epub ahead of print)
12. Siepen, J.A., et al.: An informatic pipeline for the data capture and submission of quantitative proteomic data using iTRAQTM. Proteome Sci. 5, 4 (2007)
13. Perkins, D.N., et al.: Probability-based protein identification by searching sequence databases using mass spectrometry data. Electrophoresis 20(18), 3551–3567 (1999)
14. Nilse, L., et al.: SILACAnalyzer - a tool for differential quantitation of stable isotope derived data. In: CIBB, 6th International Meeting on Computational Intelligence Methods for Bioinformatics and Biostatistics, Genoa (2009)
15. Garwood, K., et al.: Model-driven user interfaces for bioinformatics data resources: regenerating the wheel as an alternative to reinventing it. BMC Bioinformatics 7, 532 (2006)
16. Swainston, N., et al.: Enzyme kinetics informatics: from instrument to browser. FEBS J. (submitted 2010)
17. Michaelis, L., Menten, M.L.: Die Kinetik der Invertinwirkung. Biochem. Z, 49, 333–369 (1913)
18. Jameson, D., et al.: Lightweight Experimental Data Management for Systems Biology (submitted 2010)
19. Degtyarenko, K., et al.: ChEBI: a database and ontology for chemical entities of biological interest. Nucleic Acids Res. 36, D344–D350 (2008)
20. Issel-Tarver, L., et al.: Saccharomyces Genome Database. Methods Enzymol. 350, 329–346 (2002)
21. Schneider, M., et al.: The UniProtKB/Swiss-Prot knowledgebase and its Plant Proteome Annotation Program. J. Proteomics 72, 567–573 (2009)
22. Li, P., et al.: Automated manipulation of systems biology models using libSBML within Taverna workflows. Bioinformatics 24, 287–289 (2008)

An Integration Architecture Designed to Deal with the Issues of Biological Scope, Scale and Complexity

Hector Rovira, Sarah Killcoyne, Ilya Shmulevich, and John Boyle

Institute for Systems Biology, Seattle, WA, 98103

Abstract. This paper discusses a general purpose software architecture, called Addama, which is used to support the rapid integration and analysis of high volumes of complex biological data. It does this by providing: adaptable software which enables interoperable data access; a step-wise and flexible integration strategy, allowing new information to be overlaid on top of existing annotations and context graphs; and through the provision of asynchronous messaging to support rapid integration of new analysis mechanisms. This work is illustrated through the Cancer Genome Atlas (TCGA) study. Addama is being used within a TCGA analysis center to identity new therapeutic intervention approaches by equating clinical outcomes with underlying genomic effects across heterogeneous data from approximately 20,000 patient samples. Addama supports projects like the TCGA through accepting that biological understanding continually changes, and that the rapid integration of new information and analyses is an essential requirement when supporting research.

Keywords: data management, integration architecture, database, schema-free.

1 Introduction

Biological data is complex, and our understanding of the underlying principles of biological systems is continually evolving. For example, a single base pair mutation may result in a large, disease related phenotypic change. The mechanisms behind these effects can vary from alternative splicing, to new gene fusions or mislocalization. While there are many other systems to consider in biological data (e.g. protein interactions, non-coding RNA, metabolics), within genomics alone the roles of factors relating to epigenetic, non-protein encoding elements, and chromosome territories are continually changing our understanding of gene function and regulation.

Complexity is just one of the problems faced in the integration of biological data. This problem is compounded by the rate at which the data is growing in both scale and scope. This change in scale has been driven by a dramatic increase in the development and use of high throughput measurement technologies, and an associated decrease in the cost of running experiments. The widespread adoption of high throughput sequencing technologies is largely responsible for the change in magnitude of the experimental data size. In addition, new proteomics and cellular imaging technologies and experiments strongly influence the manner in which data management is being designed. The third aspect to consider is the scope of experimental

P. Lambrix and G. Kemp (Eds.): DILS 2010, LNBI 6254, pp. 179–191, 2010.

investigation, with systems biology requiring simultaneous measurements of a number of different parameters using multiple technologies (e.g. FACS, proteomics, BS-Seq, Chip-Seq).

To support this change in complexity, scope and scale new data integration architectures are needed. This paper discusses one such architecture, called Addama, which has been specifically designed to be used within systems biology based investigations. Addama has been in development for three years, and is used to support a number of high profile systems biology based research projects including:

- **The Cancer Genome Atlas (TCGA):** the architecture supports the integration and analysis of data from the Cancer Genome Atlas, as it is used to directly support one of the TCGA Genome Data Analysis Centers. This data is derived from 20,000 patient samples and includes genomic, epigenomic and clinical record information.
- **Atlas of the detectable human proteome (MRM Atlas):** the architecture is being used to support a system to create a complete map of the human proteome. This map aims to analyze approximately 1 million human peptides, and provide the tools to design and undertake targeted proteomics experiments using this reference data set.
- **Systems Approach to Immunity:** the architecture supports a distributed multi site investigation into the mouse innate immune system. Each site generates different types of information (including FACS, gene expression and proteomics profiles), which are then analyzed centrally. The data is then made available to the community (www.systemsimmunology.org).

The first section in this paper gives a background to the problems of data integration in the biomedical sciences. The example section describes the TCGA and how Addama is used to support data integration and analysis. The architecture section describes how the system was designed, and the final section outlines the benefits and deficiencies within Addama.

2 Background

This section provides a brief history of enterprise integration systems developed within the life sciences and explores the reasons for their lack of adoption. Although these powerful technologies have not been widely adopted in the research community, the different strategies that have been used offer valuable lessons which have guided the development of Addama.

Integration systems have been under development for over 15 years, and they generally followed the technology trend of the time [1-9], with each of them applying the latest standards and methodologies to the problem of integrating life science data and tools.

The adoption of these solutions in the research environment has been limited for a number of reasons, including: the time, effort and expertise needed to integrate the technologies was considerable, and the cost/benefit in terms of resources was difficult to justify; as research is continually evolving due to new findings and new instrumentation, the static formalisms (e.g. ontologies, interfaces or schemas) required by most approaches were hard to establish and maintain [10]. These constraints limited the ability to extend the functionality and incorporate tools in the required ad-hoc manner. In retrospect, and obviously subjectively, different facets of the systems that helped or hindered their adoption can be identified. One important facet is the ability to integrate "as is", so that databanks and tools do not have to be continually updated to reflect changes in the integration infrastructure or data (e.g. the early success with SRS in integrating disparate databanks using a separate parsing language, the use of BioMoby to integrate different services). Conversely the problems with predefined standards in the life sciences can be illustrated through the lack of use of systems that depended upon them (e.g. Alliance, GKP), and the importance of having a lightweight approach to integration can also be seen (e.g. by comparing the relative success of caBIG against caGrid).

The infrastructure, expertise and effort required to adopt enterprise technologies are not justified to a typical researcher. Instead, scripting languages (e.g. Perl, Ruby or Python) are often used either in conjunction with file exchange formats or as part of a two tier database architecture. These simple technologies have low learning curves that allow any researcher with minimal software training to rapidly develop and deploy a solution in their working environment. This has led to the prevalence of an "organic" development model, which does not generally result in software that is reusable or extensible. It also demands a higher tolerance for bugs and must be re-built to respond to higher performance demands (e.g. large data sets, high throughput analysis). The requirement in research has always been the ability to develop solutions quickly to satisfy short term needs, the trade off is that the resulting software is difficult to propagate, maintain, or adapt to new usage.

There have been developments in enterprise technologies that now make them easier to deploy and use. These include a variety of design patterns, specifications and frameworks which are widely adopted within the software development community [11]. The main guiding principle for these developments in software architecture was to separate application concerns into a number of tiers, each focused on a particular area of functionality. Code simplification was aided by the use of Aspect Oriented Programming, which reduces code complexity by transparently providing support for crosscutting concerns such as security and transaction control. This movement resulted in more robust and scalable applications at reduced development and maintenance costs. These developments have made enterprise technologies markedly more suitable for the rapid pace of scientific research.

Due to the fast moving nature of research it is rare that the traditional static approaches to software design, which require formalization and specifications, are appropriate. As discussed above, previous solutions for the dissemination of analysis tools and data have encountered barriers to adoption due to a failure to appreciate the extremely dynamic nature of research activities. Continuously changing research requires technologies that can be rapidly reconfigured for new and unforeseen usage. Many of the software technologies needed to provide flexible systems already exist in the public domain, though they need to be extended to make them suitable to support research informatics. These extensions include ensuring integration with common protocols and designing rapid development of ad-hoc applications. It is important that these technologies can be easily accessed through well understood protocols and made available in a transparent and non-intrusive manner. The flexible enterprise technologies that have been used for the work discussed in this paper are those that can provide: interoperable services using technologies that allow for the rapid deployment of a Service Oriented Architecture (e.g. REST based [12] web services); and a hybrid approach to the use of relational databases [13] and schema-free data stores including content repositories and "nosql" solutions (e.g. CouchDB).

3 Example of Integration

The issues of life science research informatics are exemplified by problems with the ongoing large-scale cancer study being undertaken by centers across the US. With a goal of studying 20 different cancers in 20,000 patients over the next five years, the issues of scalability, robustness and performance are immediate. Each patient sample will be sequenced (10% will be full genome, and 90% will be exon capture) with the coverage estimates currently being 100x. Each of the resulting compressed BAM files will likely be in the 300GB+ size range, and as single point mutations can cause whole domain rearrangement this represents a large data management and mining problem. Additionally each of the 20,000 patient samples (with both normal/adjacent and cancer samples being required for each patient) will undergo characterization of methylation, copy number, gene expression and small-RNA expression changes. This characterization information presents additional data integration challenges.

Fig. 1. Data files from different repositories are brought into Addama either 'on demand' or cached. The data files can be organized in any number of hierarchies, and metadata attached. Data from multiple sources can be indexed and associated with items (e.g. gene EWSR1 with PathwayCommons). Any item can be dragged onto the main web clipboard for use in analyses.

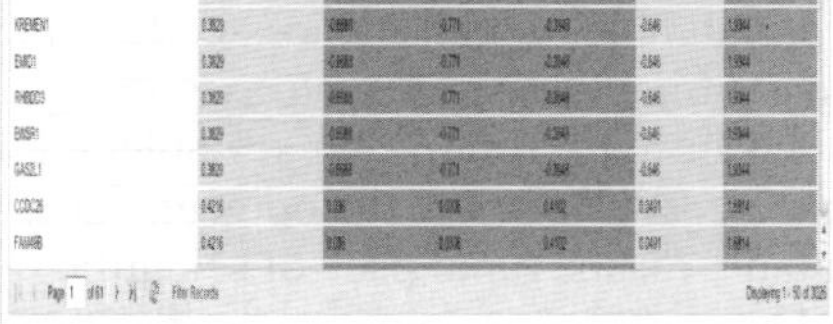

Fig. 2. A copy number variation analysis of the patient identifiers on the clipboard has been undertaken. The results are color coded by patient outcome (with red being death within three years, orange death after three years, and green survival after three years). The analysis measures the copy number variation in the different tumors. EWSR1 is shown to have high variation.

While the data integration challenges are significant, there are also problems associated with the fact that the analysis is being undertaken by distributed independent teams, and the measurement technologies are continually evolving. Addama is being used within one of the TCGA analysis centers to integrate this data and associated analyses.

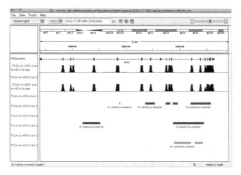
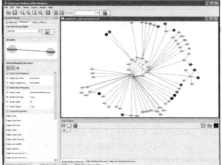

Fig. 3. By selecting an individual gene from the copy number variation table IGV (a genome browser) can be started. This automatically loads the associated patient files and as Addama integrates with the file system, further data can be loaded. In this case it is possible to see that in only those patients with poor survival is there a genome break/translocation within EWSR1.

Fig. 4. Cytoscape can also be populated with information from the Addama repository. In this case a network showing portions of the genome (and genes) that have undergone similar copy number variation is shown (and potential biomarkers are highlighted). Pathway, methylation and copy number data is also available.

An example walkthrough of the types of integration that Addama allows for is given in Figure 1-4. This walkthrough shows how Addama is being used to integrate:

- **High throughput data.** Addama is used to integrate data from two different TCGA data centers: the dbGAP (NLM) short read sequence data archive for RNASeq and Genome Assemblies, and the TCGA DCC for all other data from the TCGA genome characterization centers. Each data center has its own security models, access methods, identifiers and metadata.
- **Analysis tools.** Addama can be used to rapidly integrate analysis tools so that they are both highly accessible and can be run in an automated manner. In the walkthrough, tools for the analysis of copy number (developed by MD Anderson) and genome breaks/translocations (developed by WUSTL) are used.
- **Integration of biological knowledge.** Addama allows for knowledge to be overlaid on existing data, this is done through defining new relationships between data items or by adding additional annotations to specific data items. In the example, information from two external data sources has been overlaid: information from Pathway Commons (hosted by a second TCGA Analyses Center) is used to discover how specific pathways may have been disrupted by genomic changes, and information from MRMAtlas (hosted by the ISB) is used to identify which of

the proteins that result from putative genomic mutations in the cancers may actually be detectable (e.g. so they can act as biomarkers).

- **Integration of applications.** There are a large number of custom applications already developed by the biomedical community, these are widely used as the applications offer both useful functionality and (more importantly) are familiar to researchers. Integration with these applications is a relatively simple issue, as it involves delivery of data in the correct file format, using established protocols and providing a mechanism for triggering the application.

The purpose of Addama is to support rapid integration and analysis of large-scale, heterogeneous data. This functionality comes at a cost, as due to the distributed loose coupling, changes in data formats or analysis systems directly influence reproducibility of results (and system stability). Within the TCGA Center at ISB, Addama is principally used as a means for people to share their results with others, so that researchers can easily explore other people's results and discover if findings about specific genomic events have relevance to their own research. The walkthrough shown in Figures 1-6 demonstrates how the data and analyses can be loosely coupled and still provide for the means to undertake reasonably sophisticated analyses.

4   Software Architecture

Through judicial evaluation and extension of enterprise technologies, it is possible to provide the level of adaptability required by research. The Adaptive Data Management Service Architecture (Addama) aims to satisfy the needs of research in the life sciences by providing informatics infrastructure components to support the ad-hoc and unstructured nature of research [14]. The goal of Addama is to integrate and extend existing enterprise technologies to enable the rapid development of ad-hoc tools, and to provide a robust and scalable software infrastructure. Addama simplifies the usage of enterprise technologies by adopting standard interoperable protocols [15], and is focused on enabling the rapid integration of functionality, as well as the reduction of software complexity.

The design of Addama is guided both by the needs of research and by learning from the failure and success, of previously advocated integration systems. Researchers require a system that is flexible, to develop algorithms using familiar processes and programming languages. Research organizations also need to easily, and rapidly, adapt a system to suit the needs of the whole enterprise. The flexibility of Addama is assured by using a componentized architecture, and through the use of relevant standards and software practices. This architecture is able to support the growing number of experimental technologies and provide the means to deliver the data and analysis tools to a variety of environments.

For ongoing research an adaptable system that provides an integration framework for existing software technologies is required. This system must also have the properties that users require: provides universal access, supports discovery, and adapts to new technologies and usage. Addama is designed with these principles in mind, and marries the robustness of enterprise technologies with the responsiveness of organic software development. While adaptability was the most important requirement for

Addama, to be of use within a research enterprise our experience has shown that a system:

- **Must be accessible.** Accessibility to the data is required within research both to permit easy access to experiment data and also to enable collaboration. Individual researchers can securely access data and apply the latest analytical tools; small multidisciplinary teams can collaborate to rapidly address the needs of a specific research project; and large consortia can disseminate knowledge gathered from distributed collaborative studies and make them available to the community. Addama supports data accessibility by providing generalized REST interfaces for process and data management, ensuring interoperability between all the languages commonly used in biomedical research.

- **Must be easy to use.** This is one of the main barriers to adoption for any software technology. Resources and time are limited in research projects, so a system that is easy to learn is the one most likely to be adopted. A balance is needed between good interface design (e.g. positive transfer), the selection of appropriate technologies, and an understanding of the nature of experimental research. Rather than attempting to change the way scientists work, Addama seeks to adapt to their existing, and often chaotic, methods. Addama lowers the learning curve by providing simple interfaces for commonly used technologies.

- **Must support multiple deployment strategies.** The needs of each researcher change depending upon the project, and the computing environments vary depending on funding and project size (e.g. desktop tools, grid applications, large compute clusters). The system must be able to interact with all the different scales of scientific computing infrastructures, and should evolve to support new technologies as they become available. Addama offers a flexible architecture in which interoperating services can be deployed independently in multiple configurations.

- **Must be non-intrusive.** The majority of research organizations rely on existing technologies in which they have considerable investments. Organizations invest significant time and resources to develop workable solutions that are suitable to their immediate needs. Individual researchers invest time and effort to learn to use provided technologies. Resistance to new technologies is common, as their adoption requires additional effort, time and resources. To overcome this resistance Addama can be transparently adopted into an existing research infrastructure, as it does not require the replacement of existing solutions, but works in concert with existing standardized protocols and software components.

- **Must use appropriate software engineering practices.** Any data management system that is made available to the community should provide a high level of reliability and trust. Software engineers trust certain open source projects as they have a reputation for quality, and are based on established engineering practices. Building an integration system de novo would require significant effort and expertise, which is beyond that available to a research-funded project. Addama integrates and extends from reputable open technologies, resulting in a dramatic reduction of the overall development cost. The majority of the Addama development effort has been focused on establishing simple interfaces and designing a layered architecture that is based on good software design (e.g. SOA, SoC, IoC, AOP). The resulting software is robust, standardized, interoperable, and easily maintained.

Addama has been designed to meet these goals, and it does so by allowing for a combination of both the enterprise and organic software development models. It provides a software infrastructure that can offer the reliability and performance of an enterprise application, with the agility and responsiveness of organic software. The information and knowledge generated by individual researchers, and large-scale studies, can be unified and interconnected to provide seamless access to all the data through familiar interfaces. Addama can be integrated with the continually growing and evolving data generated in life science research. This data ranges from: large volumes of data generated by high-throughput instrumentation (e.g. mass spectrometry based proteomics, high throughput sequencing, automated imaging); results derived through automated analysis pipelines (e.g. using Bioconductor [16], GenePattern [17], TPP [18]); the output of a script created by a computational biologist (e.g. written in Python, Perl, Matlab, R); and the results of numerous small scale laboratory experiments (e.g. ELISA, Westerns, rtPCR). To enable this adaptable data management it is important to allow each area of research to define their own data models and data types. Addama ensures that there is no requirement to adhere to a specific methodology for data format, instead: information is served out using the most relevant protocols (e.g. WebDAV, LAN, REST, JSON); the system is designed so that the data is manipulated in the application layer, so that new data can be rapidly integrated or transformed for new applications; and any data can be added with additional annotations being used to supply the information necessary for integration or search, so there are no requirements to adhere to specific formats, ontology or data models.

4.1 Overview

Addama is designed to be both a data and process management system, and can be easily extended to integrate new functionality. It can be viewed as a "namespaced" SOA, where services are deployed within a specific domain. Addama provides common functionality: robust and persistent storage, controlled and secure access, search mechanisms, sharing and publishing. In addition to these core functionalities, the system supports the rapid introduction of new analysis tools and visualizations, written in any programming language, which are able to communicate and interoperate with existing tools. The Addama service architecture allows for the dynamic registration and deployment of interoperable services. A common set of interfaces for these services is then used so that they are accessible to applications (see Figure 5). Addama also provides a set of core services that are easily deployed within a J2EE application container.

This architecture enables the rapid deployment of domain-specific applications that provide new functionality while delegating responsibility for other concerns to existing services. These domain-specific applications can be developed in any web programming language (e.g., Perl CGI, Ruby-on-Rails, PHP), and are not required to be deployed within the J2EE application container or even within the same domain, instead their URLs are registered as authorized services. The services can provide access to data, present concepts, or manage logical processes through REST APIs. The default format that is used in the Addama REST response is JSON, as it becoming widely adopted as an alternative to XML for the representation and transfer of data. This provides the necessary flexibility as messages are human readable, easy to parse and generate, can represent data types and complex object structures, and are less verbose than XML. The dynamic nature of the object notation compels developers (both service and client) to adopt looser formalisms in the message structure. Service and client developers can independently modify their software to integrate the new business logic, rather than being forced to adopt new schemas.

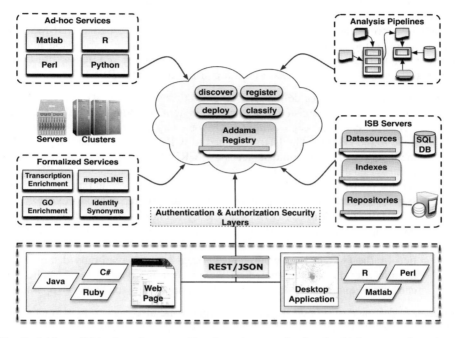

Fig. 5. Addama SOA: A registry provides the primary point for the deployment of services. Formalized services with predefined interfaces are available for specific functionality. Ad-hoc services and pipelines make use of the set of process management services, and data repositories make use of the data management services.

The core services provided with Addama are: data management services to represent and manipulate and query the data/metadata; process management services to register, access and manage the execution of analysis tools or automated pipelines.

Additionally, each of these can be overlaid with security services to ensure that users may restrict access and discovery of their data as well as providing standard mechanisms to identify users, groups and privileges within the system.

The data access services leverage existing open source software components and frameworks (e.g. Apache Tomcat, Apache Jackrabbit, Google Code, Spring Framework).

4.2 Data Management Services

As relational databases are not flexible enough to handle storing the variety of data seen in the research environment a Content Management Systems (CMS) was adopted as the underlying data storage solution. These systems are well suited to generically manage data as documents, and support dynamic annotations necessary for a flexible system. The Apache Jackrabbit implementation of the Java Content Repository (JSR-170) is used. The JCR provides for the dynamic creation of hierarchical node structures (and can be easily mapped to a file system) and properties (useful for annotations), and also provides search APIs for free-text and structured searching (e.g., XPATH).

Where appropriate, relational databases are also used in Addama. RDBMS allow for direct querying over large well defined datasets (e.g. gene expression matrices, peptide lists). The following data access services for registered JCRs, file system repositories and relational database tables are provided through REST APIs:

- **Repository Services** define URI schemes to resolve requests to data elements stored within any registered JCR or file system structure. These services (e.g. JCR, file system) assign unique URIs for every file, directory and repository node in the system. The URIs assigned to JCR nodes share mappings to HTTP methods for the common data access, or CRUD (e.g. Create, Retrieve, Update, Destroy) enabling a user to transform, and navigate the data. The file system service provides read-only access to files and directories, and has the option of serving files over HTTP or providing local file system paths to its clients.

- **SQL-DB Data Source Services** define URI schemes that resolves requests to data stored within registered databases. This service is provided as a read-only facade over database tables and views, and offers SQL-like queries as specified by the Google Data Source API. Registering a read-only database connection with the service immediately exposes it as a data source. Applications can submit queries to a data source URI by appending the "/query" operation.

Addama is designed to support the rapid integration and discovery of new functionality and data in order to ensure that researchers can quickly access new tools and integrate new types of data. In addition, these applications may participate in the propagation of generic events. Any software participating can be an event publisher or listener. There already exists a variety of specifications and enterprise software (e.g. JMS, TIBCO, ActiveMQ) to support these interactions. Addama provides services that allow these technologies to be universally accessible by offering REST APIs to:

- **Searchable Services** provides users with search capabilities over indexed data and services. The interface supports free-text searches common to the web (e.g. Google search), as well as advanced searches involving logical operations (e.g. AND, OR), annotations and concepts. It is also designed to be extensible to new search mechanisms and indexing implementations. Addama is designed to support generic indexing of annotations, keywords or contents to URIs.

- **Event Services** provides access to an event registry that facilitates the generation and propagation of events between services. The registry captures published events and guarantees their propagation to any registered client. An application may submit any custom event to this service and assume that the best effort will be made to deliver to all registered listeners. If a listener is unresponsive, the service may retry for pre-configured intervals and/or notify the user that originally registered the listener.

4.3 Process Management Services

One of the goals of the Addama architecture is to enable the rapid development and deployment of new applications and analysis tools, an essential requirement within the research environment. A variety of frameworks for the deployment of analysis scripts as web services already exist (e.g. Matlab Distributed Computing Server, Bioconductor RWebServices), but there is a need to standardize these methods to integrate any deployment and programming language. Processing services are provided in Addama to simplify the deployment, discovery and accessibility of these tools by registering them as services to handle the different processing concerns: scheduling jobs, assigning security credentials, monitoring status, deploying to distributed and parallel configurations, managing execution and reporting errors.

The process management services provide an abstraction layer between process and execution. Web service clients can monitor progress and control execution through this interface. The service provides the necessary components for execution: retrieving inputs, managing credentials, publishing status, progress and results. This allows the service to manage long running computations triggered by a user request. These jobs execute the processing logic in a parallel thread, storing the state of the execution at reasonable intervals. Developers (of clients and scripts) must understand how the asynchronous process is represented (running, completed, failed) and managed. The process management services provide the necessary infrastructure and a simple REST API to:

- Support the integration needs of process publishers and consumers. Through this interface a command-line script is registered (with search keywords) and assigned a URI. Clients can access or search for the service and use the REST API to submit jobs and monitor progress. A process specifies a set of inputs configured by the client. When new jobs are submitted for execution, the service assigns the job to a new executable thread running in a temporary workspace.

- Manage the basic functions of starting, stopping and reporting the status of an ongoing thread. This service is responsible for executing the script (business logic), capturing results and errors, and managing the state of processing threads. The scripts managed by the execution service are registered with a URI and search keywords to facilitate their discovery and may be deployed in any computer accessible through the network (e.g., desktop environment, computational cluster, Amazon EC2 cloud compute platform).

In order to reduce the amount of configuration required, the script execution service defines a simple convention that must be followed by script developers: the script receives its input parameters in the form of a standard HTTP query string (key-value pairs separated by "&" character); any type of results (e.g., images, files, annotations) to be presented to the user are stored in the directory where the script was executed; and any troubleshooting information (e.g. warnings, errors, progress) should be output to the standard output and error streams. In many cases, developers are able to deploy existing scripts by writing a simple "wrapper" that adapts to this convention. An important benefit of this approach is that it allows developers to make changes in real-time; the script can be modified and improved without having to re-deploy or even restart the service.

Each of these services is used within the TCGA project to ensure rapid access to data and integration across various types of information (e.g. sequence to pathways).

5 Conclusion

Addama is a simple service architecture that has been designed to support the demands of research, and the development of these services has been driven by the direct needs of scientists working in projects ranging in scale from individual researchers to multi-site collaborations. Addama supports scientists using heterogeneous data types (e.g. genomics, proteomics, cell imaging, FACS) and through the development of associated visualizations and analysis tools (e.g., exon expression visualization, ChIP-Seq analysis, image analysis). This work has enabled the identification of common modules and integration patterns for data collection, distribution and analysis that drive the rapid design and development of new informatics solutions.

Addama is based on a model of rapid application development, using appropriate software technologies, to address the needs of the life sciences. It defines service interfaces to integrate selected technologies with the underlying infrastructure. Thus Addama deployments can incorporate technology choices that satisfy specific requirements to support a variety of researchers using new technologies, data types and tools. Addama is used to support a number of high profile scientific computing projects spanning diverse areas of systems biology research. We are still improving the Addama system, and currently there are deficiencies in a number of areas, including: open security is required, as Addama does not integrate with the latest advances in authentication within a research enterprise (e.g. OpenSSO, OpenID) and technologies for sharing credentials across web domains (e.g. Social Networks, Hybrid Onboarding); deployment models are needed to support software applications running on cloud computing platforms or similar; and data transfer is necessary to synchronize repositories and processes for replication across multi-site collaborations.

It is important to note that Addama is not a panacea for data integration; it is an enterprise solution that has been tailored to suit the needs of ongoing research projects. There is a dire need for new approaches to informatics, as the current model of software development is not sustainable due to the rapid growth and evolution of the life sciences. This growth leads to an ever increasing demand on the informatics community to provide more complex functionality with fewer resources in shorter time periods. These new and practical approaches, such as the one discussed in this paper, will enable the development of software that is easy to: understand, so that systems can be flexibly accessed and information about data items can be conveniently updated; adapt, so that new functionality and information can be rapidly integrated; and use, so that biological data and system can be integrated and analyzed through a multitude of mechanisms. Addama can be viewed as an example of how we can solve some of the more immediate problems with life science integration by using and extending existing technologies to meet the rapidly changing requirements of research.

Acknowledgments. This work was supported by grant number P50GMO76547 from the National Institute of General Medical Sciences, NIH contract HHSN272200 700038C from the National Institute of Allergy and Infectious Diseases, grant number

U24CA143835 from the National Cancer Institute (NCI), and grant number R01-1CA1374422 from the National Cancer Institute. The content is solely the responsibility of the authors and does not necessarily represent the official views of the NIH.

References

1. Etzold, T., Argos, P.: SRS: information retrieval system for molecular biology data banks. Methods Enzymol. 266, 114–128 (1996)
2. Haas, L., et al.: DiscoveryLink: A system for integrated access to life sciences data sources. IBM Systems Journal 40(2) (2001)
3. Acero. Genomic Knowledge Platform (2002), http://www.researchobjects.com
4. Goble, C.: Putting Semantics into e-Science and Grids. In: 1st IEEE Intl. Conf. on e-Science and Grid Technologies, Melbourne (2005)
5. von Eschenbach, A., Buetow, K.: Cancer Informatics Vision: caBIG. Cancer Informatics 2, 22–24 (2006)
6. Crichton, C., et al.: Metadata-Driven Software for Clinical Trials. In: ICSE Workshop on Software Engineering and Health Care (2009)
7. Gordon, P., Trinh, Q., Sensen, C.: Semantic Web Service Provision: a Realistic Framework for Bioinformatics Programmers. Bioinformatics 23(9), 1178–1180 (2007)
8. Saltz, J., et al.: caGrid: design and implementation of the core architecture of the cancer biomedical informatics grid. Bioinformatics 22(15) (2006)
9. Microsoft. Amalga Life Sciences, http://www.microsoft.com/amalga/products/microsoftamalgalifesciences/default.mspx
10. Killcoyne, S., Boyle, J.: Chaos: Lessons Learned Developing Software in the Life Sciences. IEEE Computing in Science & Engineering 11(6), 20–29 (2009)
11. Boyle, J.: Programming Languages. In: Edwards, J.S.D., Hansen, D. (eds.) Bioinformatics Tools and Applications, pp. 403–440 (2009)
12. Fielding, R.: Architectural Styles and the Design of Network-based Software Architectures, UC-Irvine (2000)
13. Codd, E.: A Relational Model of Data for Large Shared Data Banks. Communications of the ACM 13(6), 377–387 (1970)
14. Boyle, J., et al.: Systems Biology Driven Software Design for the Research Enterprise. BMC Bioinformatics 9(295) (2008)
15. Boyle, J., et al.: Adaptable Data Management for Systems Biology Investigations. BMC Bioinformatics 10(79) (2009)
16. Gentleman, R.: R Programming for Bioinformatics 2006 (2006)
17. Reich, M., et al.: GenePattern 2.0. Nature Genetics 38(5), 500–501 (2006)
18. Nesvizhskii, A., et al.: Statistical model for identifying proteins by tandem mass spectrometry. Anal. Chem. 75, 4646–4658 (2003)
19. Shannon, P., et al.: Cytoscape: A Software Environment for Integrated Models of Biomolecular Interaction Networks. Genome Research 13(11), 2498–2504 (2003)
20. Deutsch, E., Lam, H., Aebersold, R.: PeptideAtlas: a resource for target selection for emerging targeted proteomics workflows. EMBO Rep. 9(5), 429–434 (2008)

Quality Assessment of MAGE-ML Genomic Datasets Using DescribeX

Lorena Etcheverry[1], Shahan Khatchadourian[2], and Mariano Consens[2]

[1] Instituto de Computación, Facultad de Ingeniería, Universidad de la República
lorenae@fing.edu.uy
[2] University of Toronto
shahan@cs.toronto.edu, consens@cs.toronto.edu

Abstract. The functional genomics and informatics community has made extensive microarray experimental data available online, facilitating independent evaluation of experiment conclusions and enabling researchers to access and reuse a growing body of gene expression knowledge. While there are several data-exchange standards, numerous microarray experiment datasets are published using the MAGE-ML XML schema. Assessing the quality of published experiments is a challenging task, and there is no consensus among microarray users on a framework to measure dataset quality.

In this paper, we develop techniques based on DescribeX (a summary-based visualization tool for XML) that quantitatively and qualitatively analyze MAGE-ML public collections, gaining insights about schema usage. We address specific questions such as detection of common instance patterns and coverage, precision of the experiment descriptions, and usage of controlled vocabularies. Our case study shows that DescribeX is a useful tool for the evaluation of microarray experiment data quality that enhances the understanding of the instance-level structure of MAGE-ML datasets.

1 Introduction

The collection of genes that are expressed or transcribed from genomic DNA, called transcriptome, is a major determinant of cellular phenotype and function. Differences in gene expression are responsible for both morphological and phenotypic differences as well as indicative of cellular responses to environmental stimuli. In the context of human health and treatment, gene expression measurement can help determine the causes and consequences of disease and how drugs work in cells and organisms [1].

DNAmicroarrays are devices that measure the expression of genes. Microarrays ability to measure the expression of several thousands of genes at a time has produced a quantitative change in the scale of gene measurement, leading to a qualitative change in the ability to understand regulatory processes occurring at the cellular level, and revolutionizing molecular biology and medicine [2,3]. Gene expression experiments with microarrays are complex processes with

P. Lambrix and G. Kemp (Eds.): DILS 2010, LNBI 6254, pp. 192–206, 2010.

multiple sources of variability. The functional genomics and informatics community had come together in order to develop common data exchange formats for gene expression experiments, with the goal of producing standard data products, integrating gene expression datasets, reusing previous results and allowing the independent evaluation of experiment conclusions [4]. The Microarray Gene Expression Database Society (MGED) put forward a proposal in 2001 for and experimental annotation standard, known as Minimum Information About a Microarray Experiment (MIAME), followed by an XML-based format proposal in 2004 for exchanging this information, called Microarray Gene Expression - Markup Language (MAGE-ML) [5]. Although simpler exchange formats have been proposed, such as MageTAB [6,7] and MINiML [8], MAGE-ML is still used and several gigabytes of experiments are publicly available in this format in experiments databases such as ArrayExpress [9], caArray [10], CIBEX [11] and SMD [12].

While XML provides flexibility for data providers to define their own attributes, it is also responsible for heterogeneity in data from different research groups. Differences in schema usage patterns, for example the use of optional parts of the schema, may lead to better quality levels in certain data sources. Visual exploration of the schema usage may be very helpful, since it reveals the actual structure of a collection at the instance level, element usage frequency, consistency and general patterns of usage. Understanding how the schema is used and populated helps the scientific community answer questions which are not possible by knowledge of the schema alone or by data browsing.

Even though public databases use standard exchange formats [13,14], most of them do not asses the quality of the published data [9]. Incomplete or unprecise experimental descriptions endangers the reuse of published experiments. In particular, the experimental design and experimental metadata are crucial in the selection process [15,16,17]. Since microarray experiments are complex processes, there is agreement on the importance of experimental metadata in order to allow reuse and independent evaluation of conclusions [18,19,20]. The existence of quality experimental metadata is crucial for scientifics to decide on the reproducibility and veracity of published results. For instance, geneticists would like to choose data sources with high levels of accuracy in descriptions of treatments and protocols used in the experiment.

Data quality assessment of gene expression experiments is a challenging task since the community of microarray users has not yet agreed on a framework to measure quality in microarray experiments. Without universally accepted methods for quality assessment, and guidelines for acceptance, statistics-based judgements about data quality may be perceived as arbitrary. User expectations with respect to the level of gene expression data quality vary substantially. Quality levels can depend on time frame and financial constraints associated with experimental effort, as well as the purpose of the data collection [19].

In this study we report on the application of DescribeX to explore public gene expression data collections, and gain insights of their quality and the usage of the MAGE-ML standard. Although there are many ways one can explore these data sources using DescribeX, here we focus on three particular tasks that are specific to MAGE-ML quality. The three tasks are: (a) the popularity of experimental metadata in the experiments collection (b) the use of controlled vocabulary within the collection and (c) detailed exploration of relevant parts of the experiment, the *Experiment* subtree in particular.

Our work makes three contributions. First, we show that visualization techniques and quantitative schema usage analysis supported by DescribeX can be used to explore data quality in a collection. Second, we generate specific insights into MAGE-ML standard usage and quality of the published experiment collection. Third, we propose a simple and flexible method to integrate existing data quality evaluation tools to DescribeX, extending its analysis capabilites. This work also offers a new approach and tool to the microarray community for managing, monitoring, and growing the MAGE-ML standard.

The rest of the paper is organized as follows. Next, in Section 2, we review existing literature in data quality assesment of semi-structured biological data. In Section 3 we present key aspects in MAGE-ML data collections, and how DescribeX can be used to explore its quality. Later, in Section 4 we present the application of DescribeX to a case study, analyzing the obtained results, followed by the conclusion in Section 5.

2   Related Work

Several works on biological and biomedical data quality assessment exist. In [21], production of genome data is analyzed and several error types are categorized. In [22], data quality measures are defined over sequence databases such as GeneBank and EMBL. The Qurator project [23] proposes a generic data quality framework based on users quality views, and its application to microarray experiments. However, the Qurator project proposal focuses on capturing user notion of quality rather than proposing microarray experiments quality metrics.

Data quality research has mostly focused on structured-data quality. Techniques for managing and improving data quality in semi-structured and unstructured formats are needed, as stated in [24,25]. Dataguides were introduced to help understand the structure present in semi-structured data collections [26].

Visual exploration could be used to gain insights from large scientific datasources [27]. These kinds of analysis in large data collections are non trivial and can be helped by summarization, which is not commonly supported by conventional XML tools. Existing XML browsers and tools (such as Altova XMLSpy or Stylus Studio) either cannot answer the user's questions due to maximum file size limitations, or require extensive effort to express and evaluate basic user questions. For instance, an XPath query may be created from examining an XSD schema but may not retrieve any data because the path does not actually appear in the collection.

In particular, DescribeX [28,29] has been used to explore schema usage in protein-protein interaction XML datasets [30], but not as a data quality exploration tool. In that work Samavi et al. explored the patterns of usage of the PSI-MI schema and its evolution.

3   Background

In Section 3.1 we provide an overview of the MAGE-ML data sources explored. In Section 3.2 we provide an overview of DescribeX and in Section 3.3 we expand on the application of DecribeX to explore the quality of MAGE-ML collections.

3.1   Data Sources

MAGE-ML schema is organized in several packages which represent the information described in MIAME standard [31]. A MAGE-ML document contains data regarding:

1. **experimental design**, which contains experimental objectives and a description of the experimental process
2. **arrays physical design**, including which genes are present in each array, their position within the array, etc.
3. **samples**, including data about which cells have been used, which treatments have been applied to them, etc.
4. **hybridization**, including experimental conditions such as temperature, humidity, etc.
5. **normalization techniques** used to combine data from different chips.
6. **measures**, which describes experimental results such as images, raw data, normalized data, etc.

Elements in the MAGE-ML schema are defined either as optional or mandatory. The MAGE-ML designers limit the number of mandatory elements to a minimum in order to encourage the data providers to make flexible use of the standard. NCBI's complementary effort, MINiML [32], is a data exchange format that assumes only very basic relations between the components of an experiment. Interestingly, most of the minimum required information is considered optional in the MAGE-ML standard. For example: the subtree /MAGE-ML /Experiment_package /Experiment _assnlist /Experiment in the MAGE-ML standard contains the description and annotation of the experiments overall design, including *QualityControl* information that should be reported by the scientists. Whereas most of the elements in this subtree are optional, the information they report is crucial for the interpretation and validation of the experimental results.

In this work, the data quality of 40 different datasets from the ArrayExpress repository [9] are examined. The datasets are provided by different laboratories that examine *Homo sapiens* using Affymetrix *HG-U133A* microarrays and represent almost 5% of the publicly available experiments stored in ArrayExpress that use that particular chip over Homo Sapiens. In this study, in order to minimize bias caused by data providers preferences, experiments have been carefully selected to assure that they have been performed by different laboratories and have been published in different years. A sample fragment from the collection (showing the *Experiment* element) appears in Figure 1.

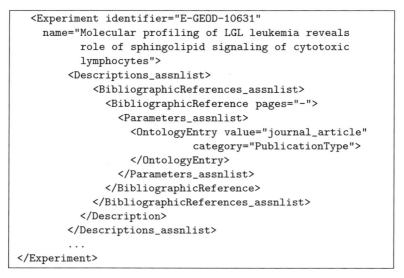

```
<Experiment identifier="E-GEOD-10631"
    name="Molecular profiling of LGL leukemia reveals
        role of sphingolipid signaling of cytotoxic
        lymphocytes">
        <Descriptions_assnlist>
            <BibliographicReferences_assnlist>
                <BibliographicReference pages="-">
                    <Parameters_assnlist>
                        <OntologyEntry value="journal_article"
                                category="PublicationType">
                        </OntologyEntry>
                    </Parameters_assnlist>
                </BibliographicReference>
            </BibliographicReferences_assnlist>
        </Description>
        </Descriptions_assnlist>
        ...
</Experiment>
```

Fig. 1. Extract of a MAGE-ML experiment

3.2 DescribeX

In this subsection, we describe how structural summaries are used to report the structure of microarray experiments that conform to the MAGE-ML schema, as well as how coverage can be used to measure the relevance of particular structures that actually occur within the collection.

The MAGE-ML schema can describe a microarray experiment's structure but, since it allows optional elements, the schema is prescriptive and may not reflect the experiment's actual structure. One way to gain insight into the actual structure of an XML document collection is by creating a structural summary of the collection. A *structural summary* describes the structure of a collection by grouping XML elements that are considered equivalent. This has the advantage that summaries can be constructed even when DTDs and XML Schemas are not present, and easily reports how optional elements (such as within the MAGE-ML schema) have been used within the collection.

DescribeX is an application to construct and explore XML structural summaries [33,28]. A structural summary is constructed by grouping elements that have equivalent neighborhoods. The neighborhood is specified using an Axis Path Regular Expression (AxPRE), which is a connected subgraph originating from each XML element in the XML graph obtained by traversing XPath axes. For example, a P* summary groups elements that have the same label path to the root, i.e., obtained by traversing the *(p)arent* axes from the element to the root. Since each element has a label path, placing each element in the XML graph into the group that has the same label path partitions all the elements in the collection.

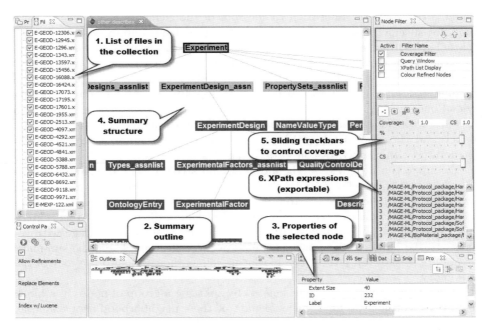

Fig. 2. DescribeX screenshot

The screenshot of DescribeX in Figure 2 depicts a P* summary of a document conforming to the MAGE-ML schema; the various views shown are now described.

View 1 is a selectable list of XML files to be summarized. View 2 shows a birds-eye view of the entire summary. View 3 shows the properties of the currently selected block. View 4 shows the structural summary. View 5 shows the sliding trackbars that control coverage. View 6 shows each block's respective XPath expression. A zoom on View 2 is depicted in Figure 3, giving a perspective of the size of the data collection.

A DescribeX structural summary (as in View 4 of Figure 2) is visualized as a graph with labeled edges and labeled nodes (referred to as *blocks*), and each block's *extent* contains the elements that have been grouped together. A block's

Fig. 3. Birds-eye view of the entire summary in DescribeX (Outline tab)

label is the common label path to the root of the elements in its block extent but, for succinctness, each block only shows the element from where the label path begins, and each block's full label path to the root is accessible either by tracing its parent path to the root or via the string in its properties view (View 3 in Figure 2). The currently selected block is *Experiment* (the block is highlighted with light grey) and has extent size 40 (as visible in the properties view), meaning that there are 40 *Experiment* elements with the same label path to the root. The view also shows each block's unique integer identifier, display label, and the label path to the root. Next, we describe how to find relevant structures within an XML collection using DescribeX.

Coverage is a way to show the most relevant structures in an instance collection by hiding non-relevant portions from its summary. Coverage can be applied at a range of values from 0% to 100%. First, blocks are ranked in order of decreasing extent size, then the top blocks are picked until the sum of their extent sizes is equal to or just greater than the selected coverage percentage with respect to the number of instance elements in the collection. Then, the summary is update to show only these top blocks as well as their paths to the root of the summary.

A structural summary can be *refined* by splitting a group of nodes in a block extent by considering an additional part of their neighborhood. For example, refining a block in a P* summary using the the AxPRE P*C will split the elements in the block's extent into groups such that each element in a group has child edges to the same set of element labels. That is, a block whose extent contains nodes that have the same path to the root are then refined according to each node's set of child elements. We show in Section 4.3 how refinement is used to reveal further information about a collection's instance-level structure.

3.3   Exploring MAGE-ML Collections

While the collection described in Section 3.1 is MAGE-ML compliant, the actual structure contains variations due to factors such as: usage of optional elements, inclusion of extra attributes and usage of ontologies as controlled vocabularies. Different groups of users, including software developers, standard designers, and scientific users, need to understand and asses the data quality of published experiments in order to either pose a query against a data source or find a best match amongst data from different sources.

DescribeX supports parsing large collections of XML-based datasets and viewing their corresponding AxPRE-based summary structure, all in a matter of minutes. For example, creating a P* summary for a 330 MB collection took 2 minutes on a conventional PC (an AMD Dual Core 2.6GHZ with 4GB RAM).

DescribeX provides several metrics that may help the user to asses the data quality of a data collection. For example, by interactively changing the *coverage*, a user can find the popularity of an element in a collection. Other metrics provided by DescribeX, such as *instance-oriented breadth and depth* [34], may also be used to gain insight of the completeness and precision of MAGE-ML data collections.

The aforesaid characteristics encourage the use of DescribeX with MAGE-ML XML collections as a way to interactively asses their data quality.

4   Results and Discussion

In this section we report the results of using DescribeX to explore and asses data quality of MAGE-ML collections. In Section 4.1, popularity and use of experimental metadata in the collection is presented. In Section 4.2 we present a simple method to extend DescribeX capabilities and its application to the measurement of accuracy in the use of controlled vocabulary. In Section 4.3 we show how DescribeX refinement capabilities can be used to perform detailed exploration of experimental datasets.

4.1   Popularity of Experimental Metadata

As stated in Section 3.1 most elements in the MAGE-ML schema that represent experimental metadata are optional. In order to explore the popularity of experimental metadata we use DescribeX to process the structure of the selected experiments, in particular the *Experiment* subtree. For the whole collection, we increase coverage at 10% intervals and measure the following four parameters:

- #T.Node: total number of instance elements that appear in the summary in a coverage interval.
- #E.Node: number of instance elements in the *Experiment* subtree in a coverage interval.
- Breadth: instance-oriented breadth for the *Experiment* subtree.
- Depth: instance-oriented depth for the *Experiment* subtree.

Figure 4 shows how relevant structure changes as coverage increases. Nodes included in coverage are dark grey, while light grey nodes represent *Experiment* subtree ancestor, root node and first level of its descendants. At each step, we characterize the extent size of the summary (#T.Node) and the extent size of the block containing all *Experiment* elements (#E.Node). There is a single block because the *Experiment* element can only appear in the path */MAGE-ML /Experiment_package /Experiment _assnlist /Experiment*

These measurements are shown in Table 1, where each column represents coverage intervals ranging form 10% to 100%. Table 1 shows the total number of nodes in each coverage interval, the number of nodes for the *Experiment* subtree compared to the total number of nodes in the summary and instance-level breadth and depth of the *Experiment* subtree. From the 782 elements available in the MAGE-ML specification, at 100% coverage we found 437 different elements (aprox. 55% of the elements available in MAGE-ML) with a total extent size of 4085989 instances. This could be explained by the fact that MAGE-ML specification has been designed to describe different kinds of experiments, and maybe several elements do not apply to the collection analyzed in this study.

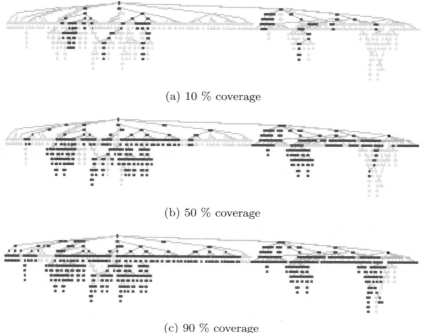

(a) 10 % coverage

(b) 50 % coverage

(c) 90 % coverage

Fig. 4. Changes in the structure as coverage increases

Table 1 also contains the number of nodes for the *Experiment* subtree compared to the total number of nodes. From the 208 elements available in the *Experiment* subtree in the MAGE-ML standard, at 100% coverage we only found 57 nodes (aprox. 25%). The absence of this information endangers reuse of experiment results, because poor quality metadata makes data meaningless [18]. Table 1 also shows instance-level breadth and depth. These two metrics help to understand if, for instance, optional nodes are used to provide more nested type information for one element (higher depth) or variety of information for different elements in a select subtree (higher breadth). In this case these two metrics indicate that the variation in breadth and depth, with respect to coverage intervals,

Table 1. Measuring popularity of experimental metadata

%COVERAGE	10	20	30	40	50	60	70	80	90	100
#T.NODE	82	129	168	209	262	285	310	352	393	437
#E.NODE	0	4	7	7	18	21	25	32	40	57
Breadth	0	2	4	5	7	9	10	12	14	17
Depth	0	2	2	2	8	8	8	10	12	12

is almost the same. This could be seen as an improvement in completeness and precision of the information.

4.2 Use of Controlled Vocabulary

While MAGE-ML provides a mechanism to standardize data representation for data exchange, a common terminology for data annotation is needed to support these standards. The MGED Ontology [35] provides terms for annotating all aspects of a microarray experiment from experiment design of the experiment and array layout, through to preparation of biological samples and the protocols used to hybridize the RNA and analyze the data. An example is depicted in Figure 5, which represents an extract of MAGE-ML description file of experiment E-MEXP-641 [1] obtained from ArrayExpress database. In this example a term in the ontology is used to specify which kind of quality control activities have been performed in the experiment. The inclusion of the *biological_replicate* individual of the *QualityControlDescriptionType* class states that biological replicates were used in the experiment.

```
<QualityControlDescription_assn>
  <Description text="RNA from 3 independent samples was
  pooled and used as template to generate double-stranded
  cDNAs">
  <Annotations_assnlist>
    <OntologyEntry value="biological_replicate"
      category="QualityControlDescriptionType">
    </OntologyEntry>
  </Annotations_assnlist>
  </Description>
</QualityControlDescription_assn>
```

Fig. 5. Example of use of *OntologyEntry* elements

Within the MAGE-ML standard, *OntologyEntry* elements are aimed to contain references to individuals in MGED Ontology, but sometimes their references are incorrect. We have previously developed an XSLT based tool which checks the syntactic correctness of these references. The tool simply checks if the pair (class, individual) contained in each *OntologyEntry* is actually a valid individual of the referenced class within the OWL file that contains the ontology.

To reuse the tool and combine it with DescribeX we have developed a simple method based on XML transformations. This method replaces each *OntologyEntry* element with new elements according to the quality evaluation result. In this case each element has been replaced by an *OntologyEntryCorrect* or an *OntologyEntryIncorrect* element. This approach represents a simple and flexible method

---

[1] http://www.ebi.ac.uk/microarray-as/ae/browse.html?keywords=e-mexp-641

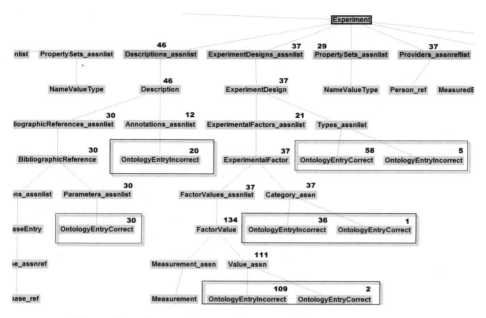

Fig. 6. Use of controlled vocabulary in the *Experiment* subtree

to integrate existent quality evaluation tools to DescribeX, since new elements may represent quality ranges (ex: *LowQuality, MediumQuality, HighQuality*).

Figure 6 shows the usage of *OntologyEntry* elements in the *Experiment* subtree and Table 2 shows the measurement results. In some subtrees correct ontology references are much more frequent than incorrect ones (black squares), whereas in others the opposite situation is found (grey squares).

Table 2. Percentage of valid an invalid ontology references in *Experiment* subtree

	Valid	Invalid
BibliographicReference/Parameters_assnlist/	100%	0%
ExperimentDesign/Types_assnlist/	92%	8%
QualityControlDescription_assn/Description/Annotations_assnlist/	75%	25%
ExperimentalFactor/Category_assn/	3%	97%
FactorValue/Value_assn/	2%	98%
Description/Annotations_assnlist/	0%	100%

4.3   Detailed Exploration of Experimental Datasets

In this section we illustrate how the explorative nature of DescribeX helps in identifying relevant parts of the experiment description. In the current MAGE-ML XML format the path */MAGE-ML /Protocol-package /Protocol-assnlist /Protocol* is the root of a subtree which has more than 20 different paths to

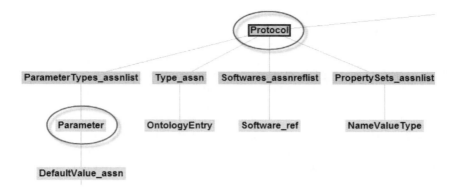

Fig. 7. Partial view of DescribeX, showing P* summary structure for *Protocol*

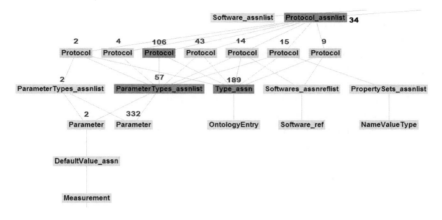

Fig. 8. Summary refinement into AxPRE P*C for *Protocol, ParameterTypes-assnlist* and *Parameter*

its leaves. Scientific users may be interested in exploring which paths are present in a particular collection. Figure 7 shows a partial view of the summary graph for the experiment collection. Although this graph is quite helpful in understanding the actual structure at the instance level (compared to the information captured from schema itself) it cannot provide enough information to reduce the number of possible queries for this element.

As presented in Section 3.2, DescribeX summaries can be refined by splitting a group of nodes in a block extent considering an additional part of their neighborhood. Using this feature, a block whose extent contains nodes that have the same path to the root, could be refined according to each node's set of child elements. In Figure 8 we show the resulting tree, where elements *Protocol, ParameterTypes_assnlist* and *Parameter* are iteratively refined. In the first iteration, refinement of the block labeled *Protocol*, splits it into seven blocks, implying the existence of seven unique substructures. In the second iteration, refinement of the block labeled *ParameterTypes_assnlist* splits into two blocks, and the last iteration affecting the the block labeled *Parameter*, splits it into two blocks. This last iteration reveals some of the reasons for different substructures. Almost every *Parameter* element is empty, whereas some of them contain information about the measurement they represent.

The refinement procedure shows that only some paths are present. Furthermore, the extent size shown beside each element in Figure 8 helps the user to focus on the more frequent XPath when required. For example the block labeled *Parameter* with an extent size of 332 in the left side of the image shows that, in this collection, protocol parameters are mostly specified as leaves, while the available MAGE-ML structure for the *Parameter* subtree is more complex.

5   Conclusion

In this paper, we introduced a new approach to quantitative and qualitative analysis of MAGE-ML XML collections using DescribeX. Our motivation was to asses data quality analyzing microarray experiments schema at an instance level. We showed how DescribeX helps to process and visualize large-scale collections of XML data.

Our work makes three types of contributions. First, we introduce new visualization and quantitative schema usage analysis techniques to explore community based XML collections. It this paper we focus on MAGE-ML data sources, but these techniques should be beneficial to other types of XML collections. Second, we gained specific insights into the MAGE-ML data standard usage by different data providers and the data quality of published experiments . This work offers a new approach and tool to the Gene Expression community for managing, monitoring, and growing MAGE-ML and other Gene Expression standards. Third, we propose a simple and flexible method to integrate existing quality evaluation tools to DescribeX. Future work includes exploring intra-experiments references consistency and the possibility of using DescribeX to support referential consistency exploration.

References

1. Lockhart, D.J., Winzeler, E.A.: Genomics, gene expression and dna arrays. Nature 405(6788), 827–836 (2000)
2. Kohane, I.S., Kho, A., Butte, A.J.: Microarrays for an Integrative Genomics. MIT Press, Cambridge (2002)
3. Stekel, D.: Microarray bioinformatics. Cambridge University Press, New York (2003)
4. Ball, C.A., Brazma, A., Causton, H., Chervitz, S., Edgar, R., Hingamp, P., Matese, J.C., Parkinson, H., Quackenbush, J., Ringwald, M., Sansone, S.A., Sherlock, G., Spellman, P., Stoeckert, C., Tateno, Y., Taylor, R., White, J., Winegarden, N.: Submission of Microarray Data to Public Repositories. PLoS Biol. 2(9) (2004)

5. Spellman, P.T., Miller, M., Stewart, J., Troup, C., Sarkans, U., Chervitz, S., Bernhart, D., Sherlock, G., Ball, C., Lepage, M., Swiatek, M., Marks, W.L., Goncalves, J., Markel, S., Iordan, D., Shojatalab, M., Pizarro, A., White, J., Hubley, R., Deutsch, E., Senger, M., Aronow, B.J., Robinson, A., Bassett, D., Stoeckert, C.J., Brazma, A.: Design and Implementation of Microarray Gene Expression Markup Language (MAGE-ML). Genome biology 3(9) (2002)
6. Rayner, T., Rocca-Serra, P., Spellman, P., Causton, H., Farne, A., Holloway, E., Irizarry, R., Liu, J., Maier, D., Miller, M., Petersen, K., Quackenbush, J., Sherlock, G., Stoeckert, C., White, J., Whetzel, P., Wymore, F., Parkinson, H., Sarkans, U., Ball, C., Brazma, A.: A Simple spreadsheet-based, MIAME-supportive Format for Microarray Data: MAGETAB. BMC Bioinformatics 7, 489 (2006)
7. Rayner, T.F., Rezwan, F.I., Lukk, M., Bradley, X.Z., Farne, A., Holloway, E., Malone, J., Williams, E., Parkinson, H.: Magetabulator, a suite of tools to support the microarray data format mage-tab. Bioinformatics 25(2), 279–280 (2009)
8. MINiML, MIAME Notation in Markup Language (2009),
   http://www.ncbi.nlm.nih.gov/geo/info/MINiML.html
9. Brazma, A., Parkinson, H., Sarkans, U., Shojatalab, M., Vilo, J., Abeygunawardena, N., Holloway, E., Kapushesky, M., Kemmeren, P., Lara, G.G., Oezcimen, A., Rocca-Serra, P., Sansone, S.A.: ArrayExpress: a Public Repository for Microarray Gene Expression Data at the EBI. Nucleic Acids Research 31(1), 68–71 (2003)
10. Bian, X., Klemm, J., Basu, A., Hadfield, J., Srinivasa, R., Parnell, T., Miller, S., Mason, W., Kokotov, D., Duncan, M., Duvall, P., Gurses, L., Boal, T., Misquitta, L., Swan, D., Wysong, R., Klink, A., Johnson, A., Fontenay, G., Liu, J., Colbert, M., Komatsoulis, G.: Data Submission and Curation for caArray, a Standard Based Microarray Data Repository System. In: Nature Proceedings (2009)
11. Ikeo, K., Ishi-i, J., Tamura, T., Gojobori, T., Tateno, Y.: CIBEX: Center for Information Biology gene EXpression database. Comptes Rendus Biologies 326(10-11), 1079–1082 (2003)
12. Demeter, J., Beauheim, C., Gollub, J., Hernandez-Boussard, T., Jin, H., Maier, D., Matese, J.C., Nitzberg, M., Wymore, F., Zachariah, Z.K., Brown, P.O., Sherlock, G., Ball, C.A.: The Stanford Microarray Database: Implementation of New Analysis Tools and Open Source Release of Software. Nucleic Acids Research 35(Database issue) (2007)
13. Gardiner-Garden, M., Littlejohn, T.: A comparison of microarray databases. Briefings in Bioinformatics 2(2), 143–158 (2001)
14. Do, H.H., Kirsten, T., Rahm, E.: Comparative Evaluation of Microarray-based Gene Expression Databases. In: BTW, pp. 482–501 (2003)
15. Canales, R.D., Luo, Y., Willey, J.C., Austermiller, B., Barbacioru, C.C., Boysen, C., Hunkapiller, K., Jensen, R.V., Knight, C.R., Lee, K.Y., Ma, Y., Maqsodi, B., Papallo, A., Peters, E.H., Poulter, K., Ruppel, P.L., Samaha, R.R., Shi, L., Yang, W., Zhang, L., Goodsaid, F.M.: Evaluation of dna microarray results with quantitative gene expression platforms. Nature Biotechnology 24(9), 1115–1122 (2006)
16. Faith, J.J., Driscoll, M.E., Fusaro, V.A., Cosgrove, E.J., Hayete, B., Juhn, F.S., Schneider, S.J., Gardner, T.S.: Many microbe microarrays database: uniformly normalized affymetrix compendia with structured experimental metadata. Nucl. Acids Res. (2007), gkm815+
17. Zeef, L.: Getting the most value out of Affymetrix array experiments (2006),
    http://nebc.nox.ac.uk/workshops/mqwshop2006.html
18. Allison, D.B., Cui, X., Page, G.P., Sabripour, M.: Microarray Data Analysis: From Disarray to Consolidation and Consensus. Nature Reviews Genetics 7(1), 55–65 (2006)

19. Brettschneider, J., Collin, F., Bolstad, B.M., Speed, T.P.: Quality Assessment for Short Oligonucleotide Microarray Data. Technometrics 50(3), 241–264 (2008)
20. Coombes, K.R., Wang, J., Abruzzo, L.V.: Monitoring the Quality of Microarray Experiments. In: Volume Methods of Microarray Data Analysis III of Biomedical and Life Sciences, pp. 25–40. Springer, US (2003)
21. Müller, H., Naumann, F.: Data quality in genome databases. In: IQ, pp. 269–284 (2003)
22. Martinez, A., Hammer, J.: Making Quality Count in Biological Data Sources. In: IQIS 2005: Proceedings of the 2nd international workshop on Information quality in information systems, pp. 16–27. ACM, New York (2005)
23. Missier, P., Embury, S.M., Greenwood, M., Preece, A.D., Jin, B.: Managing Information Quality in E-science: the Qurator Workbench. In: SIGMOD Conference, pp. 1150–1152 (2007)
24. Batini, C., Cappiello, C., Francalanci, C., Maurino, A.: Methodologies for Data Quality Assessment and Improvement. ACM Comput. Surv. 41(3), 1–52 (2009)
25. Madnick, S.E., Wang, R.Y., Lee, Y.W., Zhu, H.: Overview and Framework for Data and Information Quality Research. J. Data and Information Quality 1(1), 1–22 (2009)
26. Goldman, R., Widom, J.: DataGuides: Enabling Query Formulation and Optimization in Semistructured Databases. In: Jarke, M., Carey, M.J., Dittrich, K.R., Lochovsky, F.H., Loucopoulos, P., Jeusfeld, M.A. (eds.) VLDB 1997, Proceedings of 23rd International Conference on Very Large Data Bases, Athens, Greece, August 25-29, pp. 436–445. Morgan Kaufmann, San Francisco (1997)
27. Gray, J., Liu, D., Santisteban, M., Szalay, A., DeWitt, D., Heber, G.: Scientific Data Management in the Coming Decade. SIGMOD Rec. 34(4), 34–41 (2005)
28. Consens, M.P., Rizzolo, F., Vaisman, A.A.: AxPRE Summaries: Exploring the (Semi-) Structure of XML Web Collections. In: ICDE, pp. 1519–1521 (2008)
29. Ali, M.S., Consens, M.P., Khatchadourian, S., Rizzolo, F.: DescribeX: Interacting with AxPRE Summaries. In: ICDE, pp. 1540–1543 (2008)
30. Samavi, R., Consens, M., Khatchadourian, S., Topaloglou, T.: Exploring PSI-MI XML Collections Using DescribeX. Journal of Integrative Bioinformatics 4(3), 70 (2007)
31. Brazma, A., Hingamp, P., Quackenbush, J., Sherlock, G., Spellman, P., Stoeckert, C., Aach, J., Ansorge, W., Ball, C.A., Causton, H.C., Gaasterland, T., Glenisson, P., Holstege, F.C., Kim, I.F., Markowitz, V., Matese, J.C., Parkinson, H., Robinson, A., Sarkans, U., Schulze-Kremer, S., Stewart, J., Taylor, R., Vilo, J., Vingron, M.: Minimum information about a microarray experiment (miame)-toward standards for microarray data. Nature Genetics 29(4), 365–371 (2001)
32. Barrett, T., Troup, D.B., Wilhite, S.E., Ledoux, P., Rudnev, D., Evangelista, C., Kim, I.F., Soboleva, A., Tomashevsky, M., Edgar, R.: NCBI GEO: Mining Tens of Millions of Expression Profiles–Database and Tools Update. Nucleic Acids Res. 35(Database issue) (2007)
33. Ali, M., Consens, M., Rizzolo, F.: Visualizing Structural Patterns in Web Collections. In: WWW (2007)
34. Bex, G., Neven, F., Van den Bussche, J.: DTDs Versus XML Schema: A Practical Study. In: WebDB, pp. 79–84 (2004)
35. Whetzel, P.L., Parkinson, H., Causton, H.C., Fan, L., Fostel, J., Fragoso, G., Game, L., Heiskanen, M., Morrison, N., Rocca-Serra, P., Sansone, S.A., Taylor, C., White, J., Stoeckert, C.J.: The MGED Ontology: a Resource for Semantics-based Description of Microarray Experiments. Bioinformatics 22(7), 866–873 (2006)

Search Computing: Integrating Ranked Data in the Life Sciences

Marco Masseroli[1], Norman W. Paton[2], and Giorgio Ghisalberti[1]

[1] Dipartimento di Elettronica e Informazione, Politecnico di Milano,
Piazza Leonardo da Vinci 32, 20133 Milano, Italy
{masseroli,ghisalberti}@elet.polimi.it
[2] School of Computer Science, University of Manchester,
Oxford Road, Manchester M13 9PL, UK
npaton@manchester.ac.uk

Abstract. Search computing has been proposed to support the integration of the results of search engines with other data and computational resources. In essence, in search computing, search services provide ranked answers to requests, and mechanisms are provided for integrating results from multiple searches. This paper presents a case study of the use of a domain independent search computing platform for describing well known bioinformatics resources as search services, and for carrying out integrated analyses over the resulting services. In particular, this makes explicit how ranked data from sequence comparisons and from gene expression results can be integrated in a way that takes account of the ranked results from the different types of data. In so doing, the paper illustrates the use of ranking as a first class citizen for data integration in the life sciences, and identifies open issues for further investigation.

Keywords: search, bioinformatics, data integration, ranked data.

1 Introduction

Web search tools have become ubiquitous, with both generic and domain-specific search services, i.e. informatics services that provide results (often ranked) of user defined searches within data repositories. These services provide users with rapid and selective access to data from potentially huge repositories. However, individual search tools are often ineffective for use in applications in which the answer to a request involves combining results from more than one search engine. In particular, search services typically seek individual items that meet the criteria specified in a request, whereas in practice information relevant to a requirement may be spread over several resources. For example, if the user is interested in knowing which genes both encode proteins with high sequence similarity to a given protein and are significantly expressed in the same given biological condition or tissue, current practice typically involves the integration of results from three different searches (for similar proteins, protein encoding genes and gene expressions), where the individual search results are themselves likely to be ranked by some criteria. Such an integration task, taking

P. Lambrix and G. Kemp (Eds.): DILS 2010, LNBI 6254, pp. 207–214, 2010.

account of the rankings, is termed a *multi-domain search*, and may be carried out manually or by a custom program, but has not typically been supported directly by data integration platforms. Search computing [1] [2] provides a platform for expressing requests over multiple search services, such that the results of the integrated requests take account of the rankings of individual search results.

In the life sciences, many resources provide *vertical* search capabilities [3], in that they are focused on a single domain. For example, a sequence similarity program can be seen as a vertical search engine that takes as input a sequence and returns as output a collection of matching sequence records, ranked by some form of similarity score. In practice, many life science services provide ranked data as results, where the ranking may reflect a property of an algorithm (e.g. a similarity score) or of an experimental result (e.g. an expression level) [4]. Furthermore, it is often essential to combine multiple vertical search services to create multi-domain searches, where the different domain searches either refine or augment previous results. This paper complements a previous exploration of the envisaged relevance of search computing to the life science domain [4] by illustrating the application of an implemented search computing platform in a bioinformatics use case. A view to identifying the extent to which the existing platform for multi-domain search gives useful facilities for representing and integrating bioinformatics search services is also provided.

The remainder of the paper is as follows. Section 2 provides an overview of search computing. Section 3 characterizes both the properties of search services and the definition of requests that span multiple search services. Section 4 describes a case study that integrates biological sequence and gene expression services using the search computing platform. Section 5 discusses the lessons that can be learned from the case study and identifies some open issues that have been brought to the fore.

2   Search Computing

Search computing (http://www.search-computing.eu/) is a new multi-disciplinary approach building upon a wealth of related past research challenges, which include service/mediator based data integration, query generation and several variations of ranking in heterogeneous datasets [5] [6]. It provides the abstractions, methods, tools and computing systems required to express multi-domain queries and to build their answers [2]. For example, the following question can be answered using a multi-domain query: "Which drugs threat diseases that are likely to be associated with a given genetic mutation?" The multi-domain query can be *decomposed* into sub-queries (in this case: "Which drugs threat which diseases?"; "Which diseases are likely to be associated with a given genetic mutation?"); each sub-query can be mapped to a domain-expert server registered in the system (in this cases, calls to servers named "*Drug4Disease*", "*GeneticMutation2Disease*"); next it can be *analyzed* and translated into an internal format, which is then optimized, thereby yielding an efficient *plan* for query execution; plan execution is supported by an *execution engine*, which submits service calls to services through a *service invocation framework*, builds the *query results* by combining the outputs produced by service calls, computes the *global rankings* of query results, and outputs query results in an order that reflects their global ranking.

These transformation steps are shown in the bottom-left side of Figure 1; they are performed by the *query mapper*, *query analyzer*, *query planner* and *execution engine* modules, under the responsibility of a *query orchestrator* that starts query execution and collects query results. Suitable caching of results is performed to avoid multiple service calls when the same service results are used multiple times in a service composition. The figure shows that each of the four modules directly accepts user-provided input through suitable interfaces.

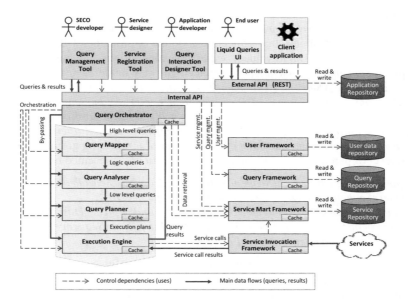

Fig. 1. Overview of the search computing framework

Services are made available to search computing through a standard format, called *service mart*, described in Chapter 9 of [2]. This is a conceptual abstraction that masks the different implementation styles of services and is tailored to the specific need to expose *search services* – i.e. services whose primary purpose is to produce ranked lists of results. These results are produced by interacting with concrete data sources, which are made available through service interfaces, wrappers, or direct access to extensional data collections (databases, excel files, and so on). To be usable for search computing, such sources must be *registered* as services in the search computing framework; this is done by defining the binding between the service mart that describes that type of resource and the operation to be invoked on the service that provides access to the resource, with its input and output parameters.

Search computing users broadly belong to two categories: *end users* can only launch predefined applications and submit input to them through forms; *expert users* may also compose queries in the context of repositories of service marts and of their composition patterns. The upper part of Figure 1 shows that end-user applications and interfaces are accessible via an external API and therefore callable from any client environment.

3   Describing Bioinformatics Search Services

A service mart models a specific type of service by describing it and its properties; each service mart definition includes a name (the service type name) and a collection of attributes (the typical input and output attributes exposed by the services of that type). Service marts have atomic attributes and repeating groups consisting of a non-empty set of sub-attributes that collectively define a property of the service mart. Atomic attributes are single-valued, while repeating groups are multi-valued. For example, a *SequenceAlignmentSearch* service mart may be defined with eight single-valued attributes (*SequenceAlignmentProgram, SearchedDB, QuerySequenceID, QuerySequence, FoundSequenceDB, FoundSequenceID, FoundSequenceSymbol* and *FoundSequenceDescription*), and one multi-valued repeating group (*Alignment*, with *Score, Expectation, Probability, MatchQuerySequence, MatchPatternSequence* and *MatchFoundSequence* sub-attributes). The schema of a repeating group is introduced by one level of parentheses, so the above example can be represented by the schema:

SequenceAlignmentSearch(SequenceAlignmentProgram, SearchedDB,
  QuerySequenceID, QuerySequence, FoundSequenceDB,
  FoundSequenceID, FoundSequenceSymbol, FoundSequenceDescription,
  Alignment(Score, Expectation, Probability, MatchQuerySequence,
  MatchPatternSequence, MatchFoundSequence))

Each service mart is associated with one or more specific *access patterns*, which abstract and logically describe the way in which data access can be effectively performed. An access pattern is a signature of the service mart in which each attribute or sub-attribute is characterized as either input (I) or output (O), depending on the role that the attribute plays in the service call. Moreover, an output attribute is designated as ranked (R) if the service produces its results in an order that depends on the value of the attribute. Access patterns can include a subset of the associated service mart attributes that are relevant for the specific data access; they can also have specific attributes not included in their service mart since not typical for the majority of services described by the service mart. For example, for the service mart *SequenceAlignmentSearch* the following access patterns can be defined:

BLAST_bySequence(SequenceAlignmentProgram^I, SearchedDB^I,
  QuerySequence^I, Email^I, FoundSequenceDB^O, FoundSequenceID^O,
  FoundSequenceSymbol^O, FoundSequenceDescription^O,
  Alignment.Score^R, Alignment.Expectation^R, Alignment.Probability^R)

BLAST_byID(SequenceAlignmentProgram^I, SearchedDB^I,
  QuerySequenceID^I, Email^I, FoundSequenceDB^O, FoundSequenceID^O,
  FoundSequenceSymbol^O, FoundSequenceDescription^O,
  Alignment.Score^R, Alignment.Expectation^R, Alignment.Probability^R)

The *SequenceAlignmentProgram* is the input attribute used to specify the sequence alignment program (e.g. *BLASTN, BLASTP*) to use in order to search, in the

SearchedDB database (e.g. *UniProtKB*), for the sequences similar to a specific query sequence; the retrieved sequences are described through the *FoundSequenceDB*, *FoundSequenceID*, *FoundSequenceSymbol* and *FoundSequenceDescription* output attributes. In the first access pattern, the query sequence is specified by providing as input its actual sequence (through the *QuerySequence* input attribute); in the second access pattern, the query sequence is specified by providing as input its ID (through the *QuerySequenceID* input attribute) in the database in which the search is performed (specified through the *SearchedDB* input attribute). In all cases, *Alignment.Score*, *Alignment.Expectation* and *Alignment.Probability* are the output attributes that can be used for providing three different rankings of the retrieved sequences and their local alignments with the query sequence, according to their similarity with the query sequence.

Each service mart is associated with one or more *service interfaces*; each of them maps an access pattern to a specific implementation and is represented as a triple including a name, a given access pattern and a service endpoint, e.g.:

> *WU-BLAST("Washington University BLAST", BLAST_bySequence,*
> *http://www.ebi.ac.uk/Tools/services/rest/wublast/run/)*

> *NCBI-BLAST("National Center for Biotechnology Information BLAST",*
> *BLAST_bySequence, http:// http://blast.ncbi.nlm.nih.gov/Blast.cgi/)*

Pair-wise coupling of service marts is defined through *connection patterns*, which completely specify the connection semantics. Every pattern has a *conceptual name* and a *logical specification*, consisting of a sequence of simple comparison predicates between pairs of attributes or sub-attributes of the two connected services; such predicates are interpreted as a conjunctive Boolean expression, and can therefore be implemented by joining the results returned by the calling service implementations. To illustrate connections, let us assume that, besides the previously defined *SequenceAlignmentSearch* service mart, a *GeneOntologyAnnotation* service mart is also available with the following schema:

> *GeneOntologyAnnotation(SequenceID, SequenceDB, GOID, GOName,*
> *AnnotationEvidence(EvidenceCode, ReferencePublication(PubMedID)))*

These service marts can be linked by a connection checking the existence of Gene Ontology annotations, with certain evidence and supporting publications, for IDs of biomolecular sequences similar to a given sequence. The connection is specified as follows:

> *ExistsGOAnnotation(SequenceAlignmentSearch, GeneOntologyAnnotation):*
> *[(SequenceAlignmentSearch.FoundSequenceDB =*
> *GeneOntologyAnnotation.SequenceDB) AND*
> *(SequenceAlignmentSearch.FoundSequenceID =*
> *GeneOntologyAnnotation.SequenceID)]*

4   Multi-domain Search over Bioinformatics Search Services

In the life sciences, numerous questions can be addressed only by comprehensively searching different types of data that are inherently ordered, or are associated with ranked confidence values. An example is the following: "Which genes encode proteins in different organisms with high sequence similarity to a given protein and are significantly expressed in the same given tissue?" By using available web services for searching bioinformatics data and taking advantage of the attributes they define for providing a ranking, search computing techniques can be applied to efficiently search for globally ranked answers to such complex questions.

The above multi-domain case study question can be decomposed into the following three single domain sub-queries: "Which proteins in different organisms have high sequence similarity to a given protein?"; "Which genes encode which proteins?"; and "Which genes are significantly expressed in the same given tissue?". Each of these sub-queries can be mapped to an available search service, i.e. a sequence similarity search program such as BLAST [7], in one of its many implementations (e.g. WU-BLAST - http://www.ebi.ac.uk/blast2/), a query service in a database of genomic and proteomic data such as our GFINDer (http://www.bioinformatics.polimi.it/GFINDer/) GPDW [8] [9], and a search engine over a repository of gene expression data such as ArrayExpress Gene Expression Atlas (http://www.ebi.ac.uk/gxa/) [10], respectively. According to the search computing framework described in the previous sections, each of these search services can be modelled with a service mart and one or more access patterns, which describe the service and its input (I), output (O) and ranked (R) attributes for the specific data accesses available. Their access patterns and pair-wise coupling connection patterns useful for computing the answer to the considered case study question are as follows:

> $WU\text{-}BLAST(SequenceAlignmentProgram^I, SearchedDB^I, QuerySequenceID^I,$
> $Email^I, FoundSequenceDB^O, FoundSequenceID^O, FoundSequenceSymbol^O,$
> $FoundSequenceDescription^O, BestAlignment.Probability^R)$

> $GPDW\_Gene2Protein(ProteinDB^I, ProteinID^I, ProteinDB^O, ProteinID^O,$
> $GeneDB^O, GeneID^O, GeneSymbol^O, Taxonomy^O)$

> $ArrayExpress(GeneSymbol^I, Organism^I, Regulation^I, Condition^I, View^I,$
> $GeneSymbol^O, Organism^O, ExperimentalFactor^O, FactorValue^O,$
> $Regulation^O, StudyNumber^R, P\text{-}value^R)$

> $ExistsCodingGene(WU\text{-}BLAST, GPDW\_Gene2Protein):$
> $[(WU\text{-}BLAST.FoundSequenceDB = GPDW\_Gene2Protein.ProteinDB$
> $AND\ WU\text{-}BLAST.FoundSequenceID = GPDW\_Gene2Protein.ProteinID)]$

> $ExistsExpressedGene(GPDW\_Gene2Protein, ArrayExpress):$
> $[(GPDW\_Gene2Protein.GeneSymbol = ArrayExpress.GeneSymbol$
> $AND\ GPDW\_Gene2Protein.Taxonomy = ArrayExpress.Organism)]$

The initial query can hence be expressed and executed in a search computing framework using a query interface such as that described in Chapter 13 of [2]. For example, in Figure 2 are depicted the global ranked results obtained for the following user specified search constraints: protein ID (*QuerySequenceID*) = "*uniprot:P26367*", expression type (*Regulation*) = "*up in*" and tissue (*Condition*) = "*liver*". The resulting genes (*PAX6* in human, *Pax6a* in rat and *Pax6* in mouse) represent the ordered list of genes that encode proteins with high sequence similarity to the *P26367* protein (human *Paired box protein Pax-6*) and are significantly *over expressed* in the *liver*.

Hence, according to the partial ranked results provided on April 8th, 2010 by the WU-BLAST, GPDW and ArrayExpress services registered in the search computing platform, they constitute the global ranked answer to the case study question that the platform can build by integrating the partial ranked results as shown in Chapter 11 of [2]. As expected, the resulting genes include the gene that encodes the input protein.

WU-BLAST			GPDW_Gene2Protein		ArrayExpress				
Similar Prot ID	Similar Prot ID Type	Expect	Gene Symbol	Organism	Factor Term	Factor Value	Diff Expr	Exp N°	P-value
P26630	Uniprot	13e-9	PAX6	Homo sapiens	organism part	liver	UP	1	1.6848E-5
P63016	Uniprot	14e-5	Pax6a	Rattus norvegicus	organism part	liver	UP	1	3.8488E-4
P63015	Uniprot	21e-6	Pax6	Mus musculus	organism part	liver	UP	3	3.1428E-2

Fig. 2. Global ranked results provided by the search computing to the case study question for the user input protein ID = "*uniprot:P26367*", expression type = "*up in*" and tissue = "*liver*". Expect: BLAST expectation; Diff Expr: gene differential expression.

5 Conclusions

This paper has provided a case study of the use of search computing for describing and composing search services in the life sciences, in order to automatically answer complex multi-domain queries. In this setting, a search service is essentially one in which the results of a request are ranked based on their fitness to the request. The bioinformatics services illustrated provide two forms of ranking that are common in the life sciences: *similarity-based ranking*, in which an algorithm provides a rank based on a computed property of the data sets being analyzed – BLAST is the best known such service, but essentially all sequence comparison or alignment algorithms provide analogous scores that give rise to ranks; and *experiment-based ranking*, in which an experimental method generates quantitative results that can be compared directly – gene expression microarray data in this context provide only one type of functional genomics data, and other quantitative techniques in proteomics, metabolomics, etc, give rise to analogous resources. As such, the paper has shown that mainstream bioinformatics resources can be described and composed using search computing constructs, where a global ranking is automatically computed by the search computing platform based on the rankings of the individual searches.

A few infrastructures, such as Taverna [11], are available to support integration of bioinformatics resources exposed as web services. However, they generally do not provide direct and/or transparent support for ranked data, which means that where ranking is considered, this must either form part of the integration application, or result from the use of analysis techniques that take ranking into account. By providing direct support for ranking as a first class citizen in data integration, search computing provides distinctive data integration features of relevance to the life sciences, where numerous ordered data types exist. The shown case study only takes into account three bioinformatics services and just one composition of them. When more services are registered in the search computing platform, they can be composed in different ways to answer complex queries and refine or augment query results. In so doing, search computing can support exploratory search and curiosity driven browsing of life science data that are difficult to perform otherwise, thus enabling ambitious data driven biological knowledge discovery and verification. Further work is required to enable more than a single mechanism for aggregating ordered data sets, since multiple global ranking mechanisms are needed to meet the variety of user requirements. For example, it may be appropriate to allow users to customize the global rankings, to reflect individual preferences on a search-by-search basis.

References

1. Braga, D., Ceri, S., Daniel, F., Martinenghi, D.: Mashing up search services. IEEE Internet Comput. 12(5), 16–23 (2008)
2. Ceri, S., Brambilla, M. (eds.): Search Computing - Challenges and Directions. LNCS, vol. 5950. Springer, Heidelberg (2010)
3. Goble, C.A., Belhajjame, K., Tanoh, F., Bhagat, J., Wolstencroft, K., Stevens, R., Pettifer, S., Nzuobontane, E., McWilliam, H., Laurent, T., Lopez, R.: BioCatalogue: a curated Web Service registry for the Life Science community. ISMB/ECCB 2009. In: Technology Track:TT40 (2009)
4. Masseroli, M., Paton, N.W., Spasic, I.: Chapter 15: Search Computing and the Life Sciences. In: Ceri, S., Brambilla, M. (eds.) Search Computing - Challenges and Directions. LNCS, vol. 5950, pp. 291–306. Springer, Heidelberg (2010)
5. Cadag, E., Louie, B., Myler, P.J., Tarczy-Hornoch, P.: Biomediator data integration and inference for functional annotation of anonymous sequences. In: Pac. Symp. Biocomput., pp. 343–354 (2007)
6. Pihur, V., Datta, S., Datta, S.: Weighted rank aggregation of cluster validation measures: a Monte Carlo cross-entropy approach. Bioinformatics 23(13), 1607–1615 (2007)
7. Altschul, S.F., Gish, W., Miller, W., Myers, E.W., Lipman, D.J.: Basic Local Alignment Search Tool. J. Mol. Biol. 215(3), 403–410 (1990)
8. Masseroli, M., Ceri, S., Campi, A.: Integration and mining of genomic annotations: Experiences and perspectives in GFINDer data warehousing. In: Paton, N.W., Missier, P., Hedeler, C. (eds.) DILS 2009. LNCS (LNBI), vol. 5647, pp. 88–95. Springer, Heidelberg (2009)
9. Masseroli, M., Ceri, S., Tettamanti, L., Campi, A., Sormani, S.: Integration of distributed heterogeneous biomolecular data to support biological discovery. In: BITS 2009: Sixth Annual Meeting Bioinformatics Italian Society, pp. 113–114. Liberodiscrivere edizioni, Genova, IT (2009)
10. Parkinson, H., Sarkans, U., Shojatalab, M., Abeygunawardena, N., Contrino, S., Coulson, R., Farne, A., Lara, G.G., Holloway, E., Kapushesky, M., Lilja, P., Mukherjee, G., Oezcimen, A., Rayner, T., Rocca-Serra, P., Sharma, A., Sansone, S., Brazma, A.: ArrayExpress - a public repository for microarray gene expression data at the EBI. Nucleic Acids Res. 33(Database issue), D553–D555 (2005)
11. Hull, D., Wolstencroft, K., Stevens, R., Goble, C., Pocock, M., Li, P., Oinn, T.: Taverna: a tool for building and running workflows of services. Nucleic Acids Res. 34(Web Server issue), W729–W732 (2006)

Author Index

Printing: Mercedes-Druck, Berlin
Binding: Stein+Lehmann, Berlin